Mathematical Aspects of Generalized Entropies and their Applications 2009

Journal of Physics: Conference Series Volume 201

Kyoto, Japan
7-9 July 2009

Editors:

Hiroki Suyari
Atsumi Ohara
Tatsuaki Wada

ISBN: 978-1-61738-276-5

Printed from e-media with permission by:

Curran Associates, Inc.
57 Morehouse Lane
Red Hook, NY 12571

Printed by Curran Associates, Inc. (2010)

Institute of Physics
Dirac House, Temple Back
Bristol BS1 6BE UK

Phone: 44 1 17 929 7481
Fax: 44 1 17 920 0979

techtracking@iop.org

Additional copies of this publication are available from:

Curran Associates, Inc.
57 Morehouse Lane
Red Hook, NY 12571 USA
Phone: 845-758-0400
Fax: 845-758-2634
Email: curran@proceedings.com
Web: www.proceedings.com

TABLE OF CONTENTS

In the recent increasing interests in power-law behaviors beyond the usual exponential ones, there have been some concrete attempts in statistical physics to generalize the standard Boltzmann-Gibbs statistics. Among such generalizations, nonextensive statistical mechanics has been well studied for about the last two decades with many modifications and refinements. The generalization has provided not only a theoretical framework but also many applications such as chaos, multi-fractal, complex systems, nonequilibrium statistical mechanics, biophysics, econophysics, information theory and so on. At the same time as the developments in the generalization of statistical mechanics, the corresponding mathematical structures have also been required and uncovered. In particular, some deep connections to mathematical sciences such as q-analysis, information geometry, information theory and quantum probability theory have been revealed recently. These results obviously indicate an existence of the generalized mathematical structure including the mathematical framework for the exponential family as a special case, but the whole structure is still unclear.

In order to make an opportunity to discuss the mathematical structure induced from generalized entropies by scientists in many fields, the international workshop 'Mathematical Aspects of Generalized Entropies and their Applications' was held on 7–9 July 2009 at Kyoto TERRSA, Kyoto, Japan. This volume is the proceedings of the workshop which consisted of 6 invited speakers, 14 oral presenters, 7 poster presenters and 63 other participants. The topics of the workshop cover the nonextensive statistical mechanics, chaos, cosmology, information geometry, divergence theory, econophysics, materials engineering, molecular dynamics and entropy theory, information theory and so on. The workshop was organized as the first attempt to discuss these mathematical aspects with leading experts in each area.

We would like to express special thanks to all the invited speakers, the contributors and the participants at the workshop. We are also grateful to RIMS (Research Institute for Mathematical Science) in Kyoto University and the Ministry of Education, Science, Sports and Culture, Grant-in-Aid for Scientific Research (B), 18300003, 2009 for their support.

Organizing Committee

Editors of the Proceedings
Hiroki Suyari (Chiba University, Japan)
Atsumi Ohara (Osaka University, Japan)
Tatsuaki Wada (Ibaraki University, Japan)

All papers published in this volume of *Journal of Physics: Conference Series* have been peer reviewed through processes administered by the proceedings Editors. Reviews were conducted by expert referees to the professional and scientific standards expected of a proceedings journal published by IOP Publishing.

Journal of Physics: Conference Series **201** (2010) 012001 doi:10.1088/1742-6596/201/1/012001

Nonadditive entropy and nonextensive statistical mechanics - Some central concepts and recent applications

Constantino Tsallis[1,2] **and Ugur Tirnakli**[3,4]

[1] Centro Brasileiro de Pesquisas Fisicas and National Institute of Science and Technology for Complex Systems, Rua Xavier Sigaud 150, 22290-180 Rio de Janeiro-RJ, Brazil
[2] Santa Fe Institute, 1399 Hyde Park Road, Santa Fe, NM 87501, USA
[3] Department of Physics, Faculty of Science, Ege University, 35100 Izmir, Turkey
[4] Division of Statistical Mechanics and Complexity, Institute of Theoretical and Applied Physics (ITAP) Kaygiseki Mevkii, 48740 Turunc, Mugla, Turkey

E-mail: `tsallis@cbpf.br, ugur.tirnakli@ege.edu.tr`

Abstract. We briefly review central concepts concerning nonextensive statistical mechanics, based on the nonadditive entropy $S_q = k\frac{1-\sum_i p_i^q}{q-1}$ $(q \in \mathcal{R};\ S_1 = -k\sum_i p_i \ln p_i)$. Among others, we focus on possible realizations of the q-generalized Central Limit Theorem, including at the edge of chaos of the logistic map, and for quasi-stationary states of many-body long-range-interacting Hamiltonian systems.

1. Introduction

1.1. Entropy

The ubiquitous concept of *energy* is associated with the *possibilities* for the configurations of a mechanical system (e.g., the eigenvalues of the Hamiltonian of a quantum system defined in a specific Hilbert space). The concept of *entropy* emerges in an even larger domain, since it can be defined for any system, mechanical or not, which admits a set of *probabilities* for its possible configurations. For instance, if we are dealing with a quantum mechanical system, the set of probabilities typically is that corresponding to the eigenvectors of the Hilbert space. Epistemologically speaking, entropy is one of the most subtle concepts in physics. Entropy and energy together constitute the basis on which statistical mechanics — one of the pillars of contemporary physics — is constructed.

The entropy, initially defined by Clausius for thermodynamics, connects the macroscopic and microscopic worlds. Its most elementary form is the logarithmic one, first introduced by Boltzmann and refined by Gibbs, von Neumann, Shannon, Jaynes and others. For a finite discrete set of probabilities $\{p_i\}$ is given by

$$S_{BG} = -k\sum_{i=1}^{W} p_i \ln p_i \quad \left(\sum_{i=1}^{W} p_i = 1\right), \tag{1}$$

where BG stands for *Boltzmann-Gibbs*. The conventional constant k is typically taken to be the Boltzmann universal constant for thermostatistical systems, or taken to be unity in information theory. For the particular case of equal probabilities, i.e., $p_i = 1/W$, $\forall i$, we have

$$S_{BG} = k \ln W \,, \tag{2}$$

carved on stone in Boltzmann grave in Vienna. Expression (1) enables the construction of a remarkably useful physical theory, referred to as *Boltzmann-Gibbs statistical mechanics*.

Many entropic forms have been introduced, and reintroduced, since Boltzmann (see [1, 2] and references therein for details). In 1988 [3], a more general form, namely

$$S_q = k \frac{1 - \sum_{i=1}^{W} p_i^q}{q-1} \quad \left(\sum_{i=1}^{W} p_i = 1;\, q \in \mathcal{R};\, S_1 = S_{BG} \right) , \tag{3}$$

was proposed as the basis for generalizing BG statistical mechanics (into a theory now known as *nonextensive statistical mechanics* [1]). For the particular case of equal probabilities, expression (3) becomes

$$S_q = k \frac{W^{1-q} - 1}{1-q} \,. \tag{4}$$

By introducing the $q-$*logarithmic* function

$$\ln_q x \equiv \frac{x^{1-q} - 1}{1-q} \quad (x > 0;\, q \in \mathcal{R};\, \ln_1 x = \ln x) \,, \tag{5}$$

expressions (3) and (4) can be respectively rewritten as follows:

$$S_q = k \sum_{i=1}^{W} p_i \ln_q \frac{1}{p_i} = -k \sum_{i=1}^{W} p_i^q \ln_q p_i \,, \tag{6}$$

and

$$S_q = k \ln_q W \,. \tag{7}$$

The axiomatics associated with this entropy have been quite explored and interesting characterizations have emerged. For details we may address the reader to [4, 5, 6, 7, 8], among others.

1.2. Additivity versus extensivity

We adopt for *entropy additivity* the definition given in Penrose's classical book [9], namely that an entropy S is said *additive* if, for any two *probabilistically independent* systems A and B, i.e., for $p_{i,j}^{A+B} = p_i^A p_j^B$, $\forall (i,j)$, we have that

$$S(A+B) = S(A) + S(B) \,, \tag{8}$$

where $S(A+B) \equiv S(\{p_{i,j}^{A+B}\})$, $S(A) \equiv S(\{p_i^A\})$, and $S(B) \equiv S(\{p_i^B\})$.

From definition (3) it is straightforward to prove that, for any two probabilistically independent systems A and B,

$$\frac{S_q(A+B)}{k} = \frac{S_q(A)}{k} + \frac{S_q(B)}{k} + (1-q) \frac{S_q(A)}{k} \frac{S_q(B)}{k} \,. \tag{9}$$

[1] The word *nonextensive* is to be associated with the fact that the total energy of long-range-interacting mechanical systems is nonextensive, in contrast with the case of short-range-interacting systems, whose total energy is *extensive* in the thermodynamical sense.

Therefore, S_{BG} is additive, whereas S_q ($q \neq 1$) is nonadditive [2].

Entropic extensivity is a concept in some sense more subtle than additivity. An entropy S of a given system constituted by N elements is said to be *extensive* if

$$0 < \lim_{N\to\infty} \frac{S(N)}{N} < \infty\,, \tag{10}$$

i.e., if $S(N) \propto N$ for $N >> 1$. We see therefore that additivity only depends on the specific mathematical connection between the macroscopic entropy functional and the probabilities of the configurations of the system. Extensivity depends on this but *also* on the specific system, more precisely on the nature of the correlations of its elements, and therefore of its collective configurations.

The distinction between additivity and extensivity has already been illustrated in simple probabilistic systems [10]. It has also been shown for the so-called *block entropy* (entropy of a subsystem of the entire system) of strongly quantum entangled fermionic and bosonic systems [11].

For the probabilistic system it has been shown [10] that

$$S_{q_{ent}}(N) \propto N \quad (N >> 1)\,, \tag{11}$$

where *ent* stands for *entropy*, N is the number of (strongly correlated) binary random variables, and

$$q_{ent} = 1 - \frac{1}{d}\,, \tag{12}$$

$d = 1, 2, 3, ...$ characterizing the width of an infinitely long strip of nonvanishing probabilities of a probability triangle asymptotically satisfying the Leibnitz rule (i.e., asymptotically scale-invariant).

For the (one-dimensional) fermionic system it has been shown that, at criticality at vanishing temperature, we have [11]

$$\lim_{N\to\infty} S_{q_{ent}}(N, L) \propto L \quad (L >> 1)\,, \tag{13}$$

where L is the number of first-neighboring spins (or analogous elements) within an infinitely long ($N \to \infty$) chain, and

$$q_{ent} = \frac{\sqrt{9 + c^2} - 3}{c}\,, \tag{14}$$

where c is the *central charge* ($c = 1/2$ for the Ising ferromagnet, and $c = 1$ for the isotropic XY ferromagnet, in the presence of a critical transverse magnetic field in both cases). We verify that q_{ent} monotonically increases from zero to one when c increases from zero to infinity.

For the (two-dimensional) bosonic system the results are qualitatively the same. However they have been established only numerically, not analytically.

The generic scenario which emerges is that, for a vast class of systems (but certainly not all), a value q_{ent} exists such that $S_{q_{ent}}(N) \propto N$ ($N \to \infty$), where N is the number of elements of the system under consideration (which might be the entire system, or only a large part of it). For standard systems, we have that $q_{ent} = 1$; for various classes of anomalous systems, we have $q_{ent} \neq 1$. The situation is depicted in Fig. 1.

[2] We remark that additivity is obtained whenever $(1-q)/k \to 0$. We see that this can occur in two different manners: $q \to 1$ for fixed k, and $k \to \infty$ for fixed q. The latter corresponds to the infinite temperature T limit, since in all thermostatistical systems T always appears in the form kT, thus having the dimension of an energy.

SYSTEMS	ENTROPY S_{BG} (additive)	ENTROPY S_q (q<1) (nonadditive)
Short-range interactions, weakly entangled blocks, etc	EXTENSIVE	NONEXTENSIVE
Long-range interactions (QSS), strongly entangled blocks, etc	NONEXTENSIVE	EXTENSIVE

Figure 1. Entropic additivity and entropic extensivity are different concepts. The BG entropy is additive, whereas S_q (for $q \neq 1$) is nonadditive. Extensivity depends on the system: for standard systems, S_{BG} is extensive, whereas it can be nonextensive for anomalous systems; S_q is the other way around.

1.3. Central limit theorems

The *Central Limit Theorem* within theory of probabilities basically states that the *sum* of a large number N of *independent* (or quasi-independent in some specific sense) random variables whose variance is *finite* converge, after appropriate centering and rescaling, to a Gaussian (i.e., $p(x) \propto e^{-\beta x^2}$). This distribution constitutes an attractor in the space of distributions, and is therefore thought to be the reason for the ubiquity of Gaussian distributions in nature. If the single distribution has a *divergent* variance instead (and also satisfies some supplementary mathematical conditions), the attractors are the celebrated Lévy distributions. The situation changes drastically if strong correlations exist among the N random variables. Depending on the nature of the correlations very many types of attractors might emerge. There is however a special class of strong correlations, referred to as *q-independence* [12], for which the attractors are q-Gaussians (i.e., $p(x) \propto e_q^{-\beta x^2}$, where the q-exponential function is defined as the inverse of the q-logarithmic one defined in Eq. (5)), if a specific generalized variance is *finite*. If this variance *diverges* instead, the attractors are the so-called (q, α)-stable distributions: see [13] for full details. The schematic description of these four theorems is presented in Fig. 2.

The physical-mathematical interpretation of the class of strong correlations named as q-independence is not yet fully ellucidated. However, it might well be that q-independence between N random variables implies (strict or asymptotic) probabilistic scale-invariance in the sense that

$$\int dx_N \, h_N(x_1, x_2, ..., x_N) \sim h_{N-1}(x_1, x_2, ..., x_{N-1}) \quad (N \to \infty) . \qquad (15)$$

Although probably necessary, this property is surely not sufficient. Indeed, (strictly or asymptotically) scale-invariant probabilistic models (with finite values for the appropriately generalized variance) have been analytically solved, some of them yielding, in the $N \to \infty$ limit, q-Gaussians, whereas other models yield distributions which numerically are amazingly close to q-Gaussians, but which definitively are *not* exactly q-Gaussians. Models that yield q-Gaussians are available in [15, 16]; models that have been proved [17] to be not exactly q-Gaussians are presented in [18, 19].

Like Gaussians, q-Gaussians also are ubiquitous [3] in natural, artificial and social systems. What could be the cause of such fact? It could very well be precisely the theorem appearing in Fig. 2 which corresponds to q-Gaussian attractors. In what follows we shall exhibit various nearly q-Gaussian distributions: in Section 2 for dissipative one-dimensional dissipative maps, in Section 3 for long-range-interacting many-body classical Hamiltonians. These two systems share a crucial property, namely that they have a maximal Lyapunov exponent which approaches zero, thus excluding strong chaos. Finite-size or finite-precision effects are present in them: we mimic this property in Section 4 with a simple mathematical model. Finally we conclude in Section 5 by mentionning various systems presented in the literature which also appear to exhibit q-Gaussians.

2. Unimodal one-dimensional dissipative maps

Let us here concentrate on a paradigmatic one-dimensional dissipative dynamical system, namely the logistic map, defined as $x_{t+1} = 1 - ax_t^2$, where a is the map parameter $(0 \leq a \leq 2)$, $-1 \leq x_t \leq 1$, and $t = 0, 1, 2, ...$ Our object of interest is the sum

$$y = \sum_{i=N_0+1}^{N_0+N} (x_i - \langle x \rangle) \qquad (16)$$

in the vicinity of chaos threshold $a_c = 1.4011..$, where N_0 is the number of transient steps (typically $N_0 >> 1$) and

$$\langle x \rangle = \frac{1}{n_{ini}} \frac{1}{N} \sum_{j=1}^{n_{ini}} \sum_{i=1}^{N} x_i^{(j)} \qquad (17)$$

is the average over a large number of N iterates as well as a large number n_{ini} of randomly chosen initial values $x_1^{(j)}$ of iterates of the map. This problem has been addressed firstly in [20] and a closer look has been given very recently in [21]. Here, we try to further clarify the study and analyse it in a more compact manner.

Generically, the problem at hand is the form of the probability distribution of the random variable given in Eq. (16). Indeed, the ordinary Central Limit Theorem (CLT) (yielding the Gaussian form) is applicable when the Lyapunov exponent is positive, but, when approaching the edge of chaos from above (i.e., for $a > a_c$), an infinite number of values of a accumulate which violate this condition. The situation becomes then quite subtle, as we shall review here. Essentially, strong correlations between the iterates of the map emerge. In [21] it is shown that the problem is much more complex than the ordinary CLT, which is the case when the map is at the chaotic regime (e.g., for $a = 2$). One needs now to be careful on how we approach the chaos threshold point (a_c) and how the number of iterations (N) to be used increases to infinity. In mathematical language, this means that two limits are to be performed simultaneously, namely, $(a - a_c) \to 0$ and $1/N \to 0$. It is argued in [21] that the limit distributions appear to be of q-Gaussian type if these two limits are performed simultaneously in the following special way (which satisfies an appropriate scaling relation). We first choose a value of a above and close to

[3] The word *ubiquitous* is here used not in the strict sense of being everywhere, but only in the loose sense of being found very frequently.

Journal of Physics: Conference Series **201** (2010) 012001 doi:10.1088/1742-6596/201/1/012001

	$q=1$ [*independent*]	$q \neq 1$ (*i.e.*, $Q \equiv 2q-1 \neq 1$) [*globally correlated*]
$\sigma_Q < \infty$ $(\alpha = 2)$	$\mathbb{F}(x) = Gaussian\ G(x)$, *with same* σ_1 *of* $f(x)$	$\mathbb{F}(x) = G_q(x) \equiv G_{(3q_1-1)/(1+q_1)}(x)$, *with same* σ_Q *of* $f(x)$ $G_q(x) \sim \begin{cases} G(x) & \text{if } \lvert x \rvert << x_c(q,2) \\ f(x) \sim C_q / \lvert x \rvert^{2/(q-1)} & \text{if } \lvert x \rvert >> x_c(q,2) \end{cases}$, *with* $\lim_{q\to 1} x_c(q,2) = \infty$
$\sigma_Q \to \infty$ $(0 < \alpha < 2)$	$\mathbb{F}(x) =$ *Levy distribution* $L_\alpha(x)$, *with same* $\lvert x \rvert \to \infty$ *behavior* $L_\alpha(x) \sim \begin{cases} G(x) & \text{if } \lvert x \rvert << x_c(1,\alpha) \\ f(x) \sim C_\alpha / \lvert x \rvert^{1+\alpha} & \text{if } \lvert x \rvert >> x_c(1,\alpha) \end{cases}$ *with* $\lim_{\alpha \to 2} x_c(1,\alpha) = \infty$	$\mathbb{F}(x) = L_{q,\alpha}$, *with same* $\lvert x \rvert \to \infty$ *asymptotic behavior* $L_{q,\alpha} \sim \begin{cases} G_{\frac{2(1-q)-\alpha(1+q)}{2(1-q)-\alpha(3-q)},\alpha}(x) \sim C^*_{q,\alpha} / \lvert x \rvert^{\frac{2(1-q)-\alpha(3-q)}{2(1-q)}} \\ \text{(intermediate regime)} \\ G_{\frac{2\alpha q-\alpha+3}{\alpha+1},2}(x) \sim C^L_{q,\alpha} / \lvert x \rvert^{(1+\alpha)/(1+\alpha q-\alpha)} \\ \text{(distant regime)} \end{cases}$

Figure 2. $N^{1/[\alpha(2-q)]}$-scaled attractors $\mathcal{F}(x)$ when summing $N \to \infty$ q-independent identical random variables with symmetric distribution $f(x)$ with Q-variance $\sigma_Q \equiv \int_{-\infty}^{\infty} dx\, x^2\, [f(x)]^Q / \int_{-\infty}^{\infty} dx\, [f(x)]^Q$ ($Q \equiv 2q-1$; $q_1 = (1+q)/(3-q)$; $q \geq 1$). *Top left:* The attractor is the Gaussian sharing with $f(x)$ the same variance σ_1 (standard CLT). *Bottom left:* The attractor is the α-stable Lévy distribution which shares with $f(x)$ the same asymptotic behavior, i.e., the coefficient C_α (Lévy-Gnedenko CLT, or α-generalization of the standard CLT). *Top right:* The attractor is the q-Gaussian which shares with $f(x)$ the same $(2q-1)$-variance, i.e., the coefficient C_q (q-generalization of the standard CLT, or q-CLT). *Bottom right:* The attractor is the (q,α)-stable distribution which shares with $f(x)$ the same asymptotic behavior, i.e., the coefficient $C^L_{q,\alpha}$ (q-generalization of the Lévy-Gnedenko CLT and α-generalization of the q-CLT). The case $\alpha < 2$, for both $q = 1$ and $q \neq 1$ (more precisely $q > 1$), further demands specific asymptotics for the attractors to be those indicated; essentially the divergent q-variance must be due to fat tails of the power-law class, excepting for possible logarithmic corrections (for the $q = 1$ case see, for instance, [14] and references therein).

a_c. We then calculate the quantity n defined as follows:

$$n = -\frac{\log |a - a_c|}{\log \delta}, \tag{18}$$

where $\delta = 4.6692011...$ is the Feigenbaum constant. We then denote by k the nearest integer value of $2n$, and define N^* through

$$N^* = 2^k. \tag{19}$$

The q-Gaussians numerically appear to gradually emerge when we choose $N = N^*$ and keep making $(a - a_c) \to 0$ (hence $N = N^* \to \infty$). Let us refer to this region as the q-Gaussian probability distribution functions (PDFs) one and analyse its borders. To do this, typical values for a have been used: they are given in Table 1, as well as their related parameters; a values can easily be taken so that $2n$ values would be obtained with the same precision. Each group with the same precision enables the construction of a linear curve in the space $1/N$ vs $(a-a_c)^s$, where $s \equiv \ln 4/\ln\delta \simeq 0.9$: see Fig. 3. Numerical inspection has shown that no other q-Gaussian linear curves occur at the left of the largest slope and at the right of the smallest slope in Fig. 3. All q-Gaussian lines appear to exist only between these two extremes; see examples in Fig 4 as well as Fig. 5 where (q, β) pairs of all studied cases are plotted.

Table 1. The values of map parameter a and N^* used in this work. The values of $2n$ (obtained from the scaling relation), and the values of q and β (estimated from simulations) are also listed. In this work, the critical value a_c is approximated as $a_c = 1.40115518909 20505$.

a	$a-a_c$	$2n$	$N^*=2^k$	q	β
1.40159888	0.00044369	10.02	2^{10}		
1.40125021721	0.00009503	12.02	2^{12}	1.70	6.6
1.40117554121	0.00002035	14.02	2^{14}		
1.40115954790	0.00000436	16.02	2^{16}		
1.40152683	0.00037164	10.25	2^{10}		
1.40123478	0.00007959	12.25	2^{12}	1.70	6.5
1.401172235	0.00001705	14.25	2^{14}		
1.4011588398	0.00000365	16.25	2^{16}		
1.4014862	0.00033101	10.40	2^{10}		
1.401226075	0.00007088	12.40	2^{12}	1.70	6.8
1.401170372	0.00001518	14.40	2^{14}		
1.401158441	0.00000325	16.40	2^{16}		
1.401464065	0.00030888	10.49	2^{10}		
1.401221341	0.00006615	12.49	2^{12}	1.68	6.7
1.4011693567	0.00001417	14.49	2^{14}		
1.4011582234	0.00003034	16.49	2^{16}		
1.40145934	0.00030415	10.51	2^{11}		
1.40122033	0.00006514	12.51	2^{13}	1.61	6.8
1.40116914	0.00001395	14.51	2^{15}		
1.401158177	0.00000299	16.51	2^{17}		
1.40138924	0.00023405	10.85	2^{11}		
1.401205317	0.00005013	12.85	2^{13}	1.65	6.5
1.401165925	0.00001074	14.85	2^{15}		
1.4011574883	0.00000230	16.85	2^{17}		
1.40136531	0.00021012	10.99	2^{11}		
1.40120019	0.00004500	12.99	2^{13}	1.63	6.5
1.401164827	0.00000964	14.99	2^{15}		
1.4011572532	0.00000206	16.99	2^{17}		

The q-Gaussian-like PDFs are lost as soon as the scaling relation (Eq. (18)) is ignored, i.e., if values for N larger or smaller that N^* are used. Two new regions emerge. If we use values for N that are sensibly smaller than N^*, or, in other words, the value of a that is being used is too close to the critical point, peaked PDF's are observed (see also [22]). On the other extreme, if we use values for N that are sensibly larger than N^*, we observe in most of the cases Gaussian

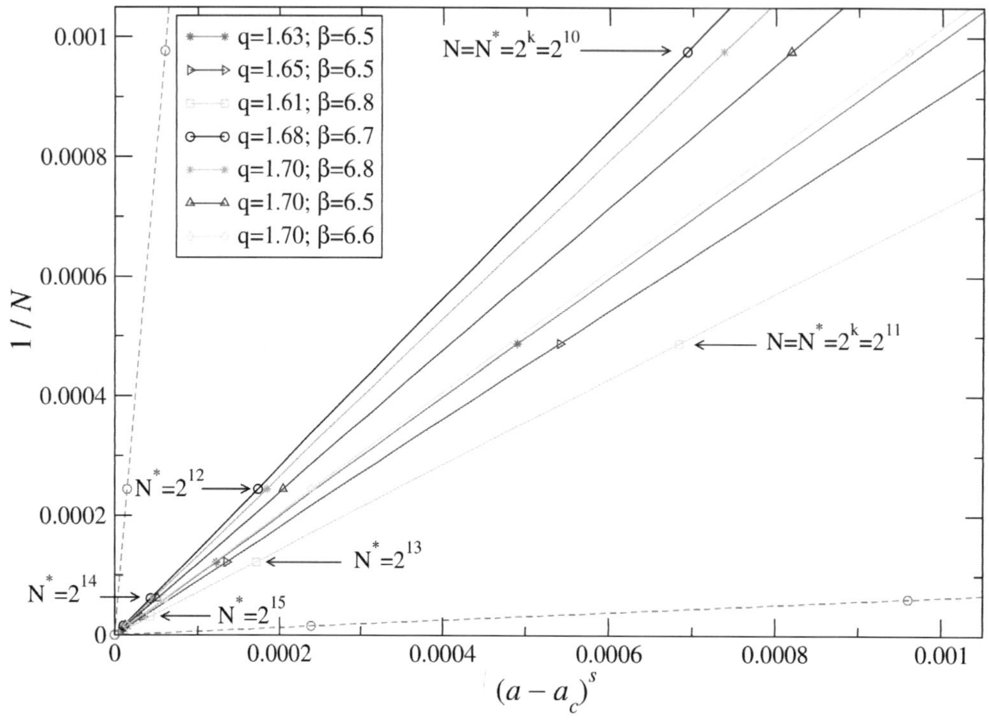

Figure 3. View of the parameter region $1/N$ vs $(a - a_c)^s$ with $s = \ln 4/\ln \delta \simeq 0.9$ coming directly from the scaling relation (18). Typical values of a are shown, and their correspondiong values of N^* are determined from the scaling relation (18). Any other possible choices for N^* yield lines that remain between the lines with the largest and the smallest slopes shown in the figure. If one approaches the critical point (origin) along any of these lines in this region, the probability distribution function (PDF) appears to gradually approach, excepting for a small oscillating contribution, a q-Gaussian. The regions at the left of the largest slope and at the right of the smallest slope are not accessible as far as N^* values are concerned. Four of the seven q-Gaussian examples are presented in Fig. 4; the peaked (almost vertical line) and the Gaussian (almost horizontal line) examples are presented in Fig. 6.

PDF's, i.e., $P(y) = e^{-y^2/(2\sigma^2)}/\sqrt{2\pi\sigma^2})$. Two representative examples for the two regions outside the q-Gaussian one are indicated with dashed lines (magenta) in Fig. 3 and illustrated in Fig. 6.

3. Long-range-interacting many-body classical Hamiltonian systems

Let us consider the following classical Hamiltonian of N interacting planar classical rotators on a d-dimensional (simple hypercubic) lattice [23]:

$$\mathcal{H} = \sum_{i=1}^{N} \frac{p_i^2}{2} + \frac{1}{2}\sum_{i,j} \frac{1 - cos(\theta_i - \theta_j)}{r_{ij}^{\alpha}} \quad (\alpha \geq 0)\,, \tag{20}$$

where r_{ij} runs over all possible distances within the d-dimensional lattice. The particular case $\alpha = 0$ is referred to in the literature as the HMF model [24], and has been intensively studied in the last decade (in its standard representation the coupling constant is divided by N, which artificially makes the total energy extensive in the thermodynamical sense). Its dynamical molecular approach has exhibited a variety of interesting phenomena: see, for instance, [1] and

Journal of Physics: Conference Series **201** (2010) 012001 doi:10.1088/1742-6596/201/1/012001

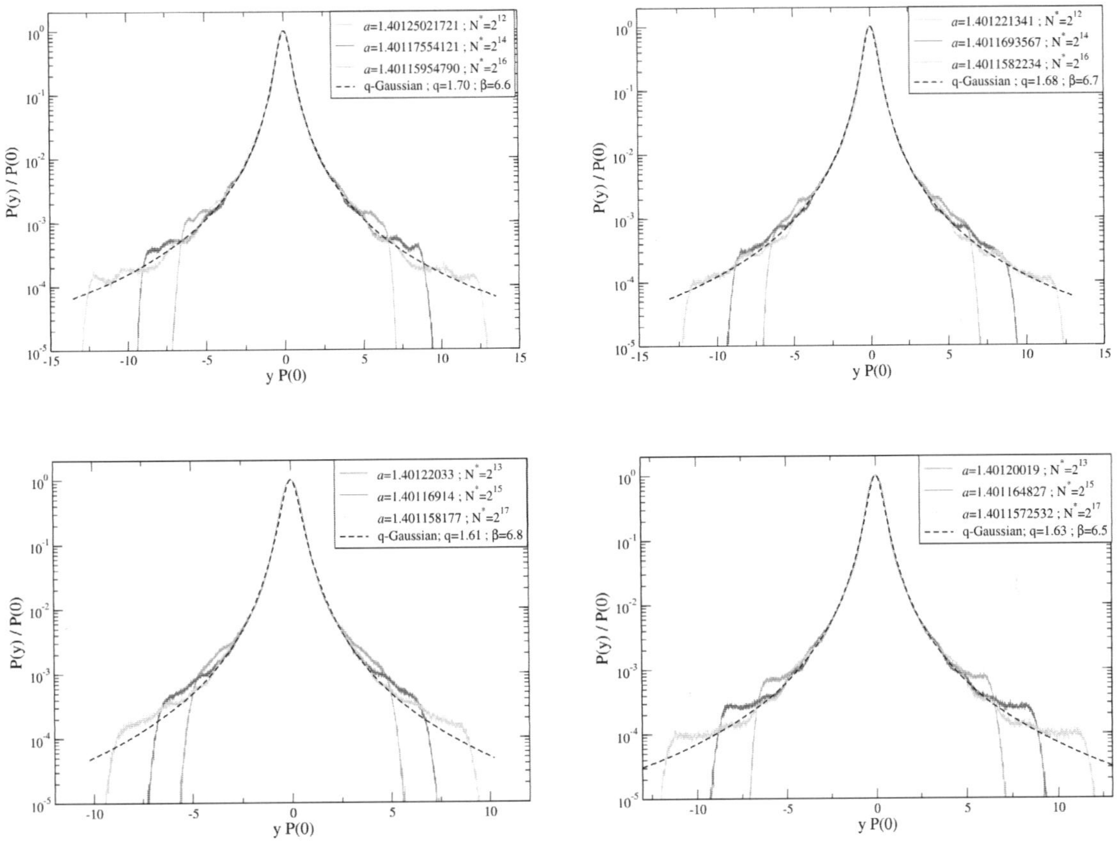

Figure 4. Data collapse of PDF's for the cases $N = N^* = 2^k$ for four of the seven representative examples given in Table 1. The (possible) q-Gaussians are approached through finite-N effects analogous to those exhibited in Fig. 9.

references therein. In particular, for the isolated system (microcanonical ensemble) at energy per particle equal to 0.69, long-standing quasi-stationary states (QSS) emerge when certain classes of initial conditions (usually called *water-bag* initial conditions) are used. Within the water-bag conditions, one may consider initial magnetization equal to zero (usually referred to as $M = 0$), or equal to its maximal value (usually referred to as $M = 1$), or values in between. Such choices influence the specific trajectory of the full system within its $2N$-dimensional phase space (Gibbs Γ space). It has been shown that, for $M = 1$ initial conditions (possibly for virtually all values of initial M), *ergodicity is broken.* Indeed the *summed* (over n equidistant instants) one-velocity marginal PDF differs when we take ensemble-average or time-average: see [25] and references therein. Many of the time-averaged PDF's numerically approach a q-Gaussian (see Fig. 7 for one such example). Although no analytical proof is available at the present time, this might be a consequence of the q-Central Limit Theorem [12], within which q-Gaussians are the attractors in the space of PDF's. The finite-size effects are illustrated in Fig. 7.

4. Simple mathematical model for crossover between q-Gaussians

As a mathematical simple model for finite-size or finite-precision effects exhibiting the gradual approach to q-Gaussians, let us consider the following differential equation [1, 26]:

$$\frac{dy}{dx} = -ay^r - by^q \quad (b \geq a \geq 0;\ q > r;\ y(0) = 1)\,. \tag{21}$$

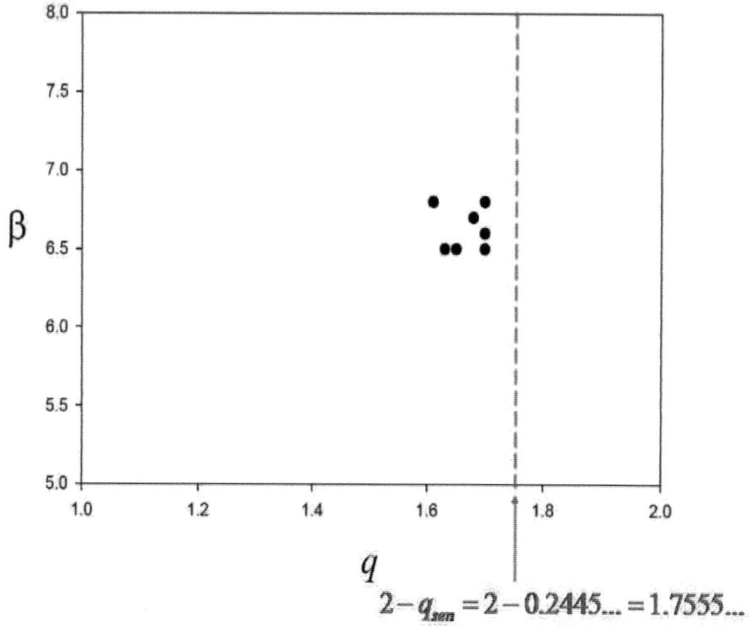

Figure 5. The parameters (q, β) corresponding to the seven q-Gaussians indicated in Fig. 3. These specific seven examples appear to exclude the value $2 - q_{sen} = 1.7555...$, which could have been a plausible result. At the present numerical precision, even if quite high, it is not possible to infer whether the analytical result corresponding to the present observations would be only one or a set of q-Gaussians, assuming that exact q-Gaussians are involved, on top of which a small oscillating component possibly exists.

If $a = 0$ the solution is

$$y = e_q^{-bx} \,. \tag{22}$$

If $b = 0$ the solution is

$$y = e_r^{-ax} \,. \tag{23}$$

If $b > a > 0$ a crossover occurs from the q-exponential solution for x not too large to the r-exponential solution for x large enough; *increasing size or increasing precision* for specific models (such as the logistic map at its edge of chaos, or the HMF Hamiltonian model at its QSS state) acts analogously to *decreasing values of* a towards the limit $a = 0$, for fixed b. Let us address this interesting case. ¿From (21) we obtain

$$\begin{aligned} x = -\int_1^y \frac{du}{au^r + bu^q} &= \frac{r}{b(-1+q)(q-r)} - \frac{\left(r + (q-r)\,{}_2F_1\left[\frac{-1+q}{q-r}, 1, 1 + \frac{-1+q}{q-r}, -\frac{a}{b}\right]\right)}{b(-1+q)(q-r)} \\ &- \frac{ry^{1-q}}{b(-1+q)(q-r)} + \frac{y^{1-q}\left(r + (q-r)\,{}_2F_1\left[\frac{-1+q}{q-r}, 1, 1 + \frac{-1+q}{q-r}, -\frac{ay^{-q+r}}{b}\right]\right)}{b(-1+q)(q-r)} \end{aligned} \tag{24}$$

where ${}_2F_1$ is the hypergeometric function. In general, this function does not admit an *explicit* expression in the form $y(x)$. An exception is the $r = 1$ case, which yields

$$y = \frac{1}{\left[(\frac{b}{a} + 1)e^{(q-1)\,a\,x} - \frac{b}{a}\right]^{\frac{1}{q-1}}} \,. \tag{25}$$

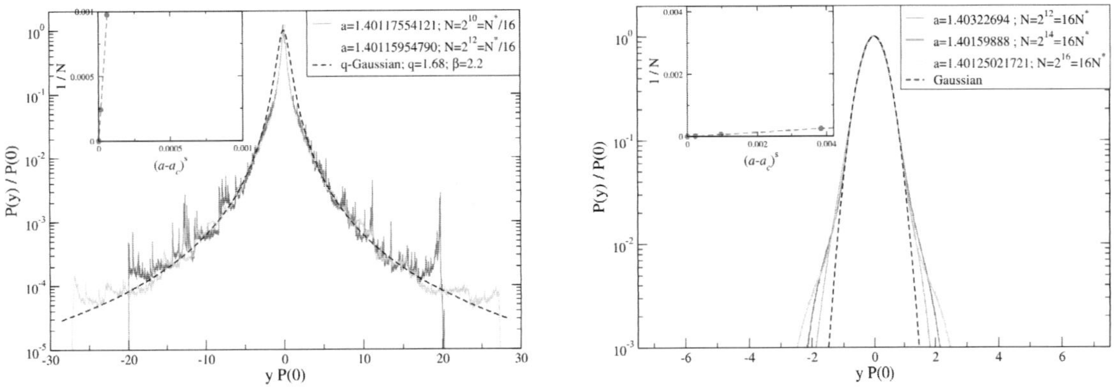

Figure 6. Representative examples from the peaked region (left) and the Gaussian region (right).

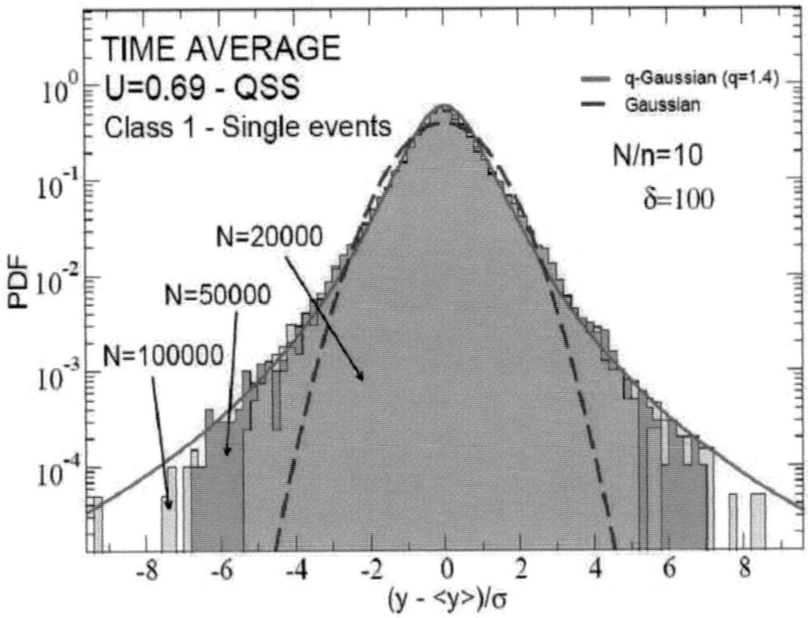

Figure 7. Velocity distribution for the HMF model at its quasi-stationary state (QSS) [Data from Fig. 2 of [25]]. The finite-N effects are visible, gradually approaching a (possible) q-Gaussian when both N and n are diverging. An analogous behavior is shown in Fig. 9. For comparison, a Gaussian is shown as well.

The $r = 0$ case must be handled through the explicit $x(y)$ form, namely

$$x = \frac{1}{a}\Big\{ {}_2F_1\Big[\frac{1}{q}, 1, 1 + \frac{1}{q}, -\frac{b}{a}\Big] - y\, {}_2F_1\Big[\frac{1}{q}, 1, 1 + \frac{1}{q}, -\frac{b}{a} y^q\Big]\Big\} . \tag{26}$$

Let us address now the case of the q-Gaussians. Following the form of Eq. (21), we consider

$$\frac{dy}{d(x^2)} = -a_r y^r - (a_q - a_r) y^q \quad (a_q \geq a_r \geq 0;\, q > r;\, y(0) = 1)\,. \tag{27}$$

If $a_r = 0$, or equivalently if $r = q$, the solution is given by the q-Gaussian $y = e_q^{-a_q x^2}$. If $a_r = a_q$, the solution is given by the r-Gaussian $y = e_r^{-a_r x^2}$. For the case $a_q > a_r > 0$ and $q > r$, we obtain a crossover between these two solutions, the $|x| \to \infty$ asymptotic one being the r-Gaussian behavior.

For $r = 1$ and $q > 1$, the solution is given by

$$y = \frac{1}{\left[1 - \frac{a_q}{a_1} + \frac{a_q}{a_1} e^{(q-1) a_1 x^2}\right]^{\frac{1}{q-1}}}\,. \tag{28}$$

The general behavior of these solutions is given in Fig. 8. It is evident from this figure that this solution is not the most appropriate one for the behavior observed in the neighborhood of the logistic map edge of chaos. We notice concomitantly that the appropriate solution for the logistic map seems to be very close to the one with $r = 0$ and $q > 1$, whose solution is given by

$$x^2 = \frac{1}{a}\left\{{}_2F_1\left[\frac{1}{q}, 1, 1 + \frac{1}{q}, -\frac{b}{a}\right] - y\, {}_2F_1\left[\frac{1}{q}, 1, 1 + \frac{1}{q}, -\frac{b}{a} y^q\right]\right\}, \tag{29}$$

with $a \equiv a_0$ and $b \equiv a_q - a_0$. Indeed, this solution seems to be a very good approximation for the behavior of PDFs obtained numerically for the logistic map. This can be seen immediately whenever the representative example of the solution given in Fig. 9 is compared to the case given in Fig. 4a.

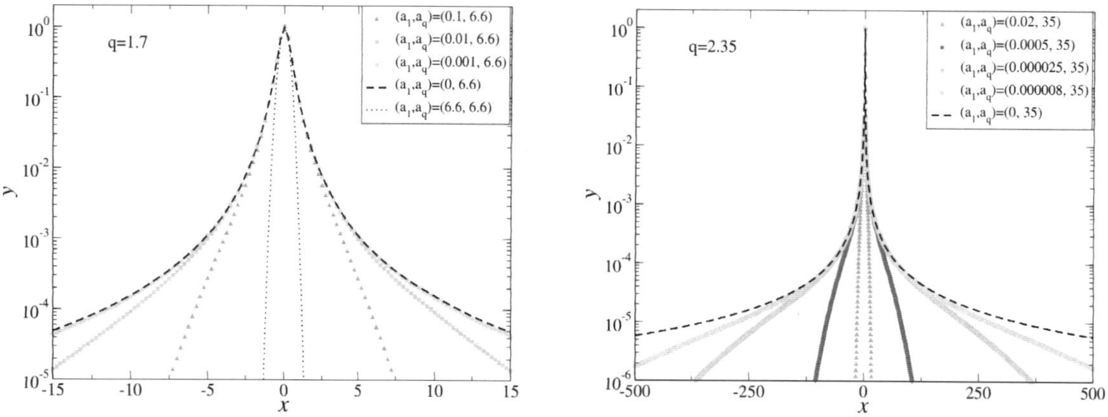

Figure 8. Crossover from q-Gaussian to successively distant Gaussians, i.e., illustrations of the case $r = 1$ and $q > 1$. *Left:* $q = 1.7$; for comparison, a Gaussian is shown as well. *Right:* $q = 2.35$.

5. Final remarks

Many q-Gaussian-like distributions have been observed in recent years in natural, artificial and social systems. Obviously, only for a mathematically formulated model, the hope exists to possibly prove analytically that the relevant distribution precisely, and not only approximatively,

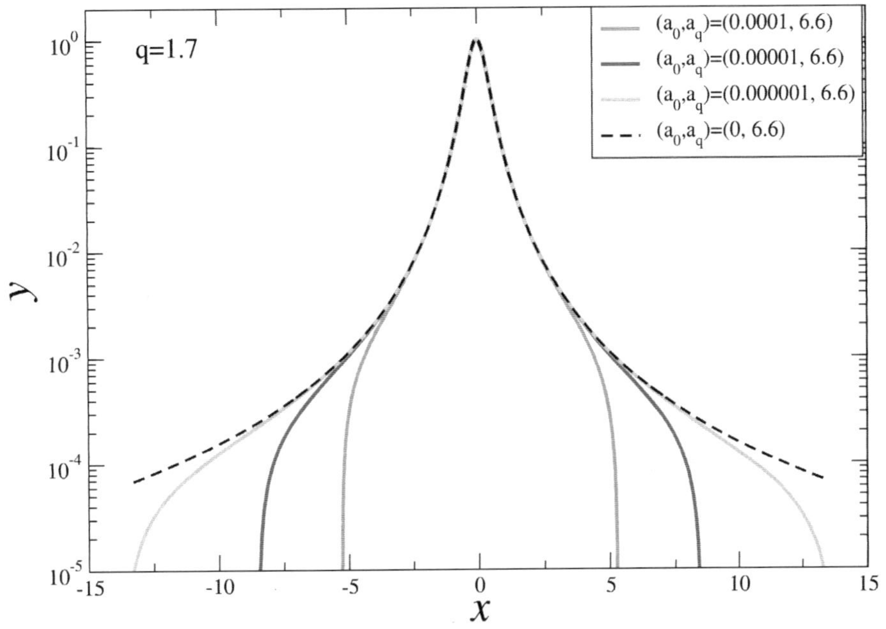

Figure 9. Crossover from 1.7-Gaussian with $\beta = 6.6$ to successively distant r-Gaussians with $r = 0$.

is a q-Gaussian. In all other cases, we can only expect for increasingly high-precision indications from real experiments or observations. Computational evidence can and does provide important hints, however never a proof.

This said, let mention in what follows some of the many other systems where q-Gaussians have been used to approach the observed PDF's: (i) The velocity distribution of (cells of) *Hydra viridissima* follows a $q = 3/2$ PDF [27]; (ii) The velocity distribution of (cells of) *Dictyostelium discoideum* follows a $q = 5/3$ PDF in the vegetative state and a $q = 2$ PDF in the starved state [28]; (iii) The velocity distribution in defect turbulence [29]; (iv) The velocity distribution of cold atoms in a dissipative optical lattice [32]; (v) Velocity distribution during silo drainage [30, 31]; (vi) The velocity distribution in a driven-dissipative 2D dusty plasma, with $q = 1.08 \pm 0.01$ and $q = 1.05 \pm 0.01$ at temperatures of $30000\,K$ and $61000\,K$ respectively [33]; (vii) The spatial (Monte Carlo) distributions of a trapped $^{136}Ba^{+}$ ion cooled by various classical buffer gases at $300\,K$ [34]; (viii) The distributions of price returns at the stock exchange [35, 36, 37]; (ix) The distributions of returns of magnetic field fluctuations in the solar wind plasma as observed in data from Voyager 1 [38] and from Voyager 2 [39]; (x) The distributions of returns of the avalanche sizes in the Ehrenfest's dog-flea model [40]; (xi) The distributions of returns of the avalanche sizes in the self-organized critical Olami-Feder-Christensen model, as well as in real earthquakes [41]; (xii) The distributions of angles in the HMF model [42]; (xiii) The distribution of stellar rotational velocities in the Pleiades [43]. Some indirect evidence is available as well: although no q-Gaussian distribution has been directly observed in some relevant physical quantity, a q-exponential relaxation has been seen in various paradigmatic spin-glass substances through neutron spin echo experiments [44].

Clearly, the simplest hypothesis which would explain the ubiquity of q-Gaussians is the validity of the q-Central Limit theorem. The involved random variables would, in such case, be expected to be q-independent [12]. The present belief is that (probabilistic) scale-invariance is necessary but not sufficient for q-independence. Further studies are needed to clarify the applicability of such ideas to the systems mentioned above, as well as possibly others.

Acknowledgments
We thank insightful remarks by C. Beck, E.M.F. Curado, A. Pluchino, A. Rapisarda, A. Robledo and B.C.C. dos Santos. This work has been partially supported by CNPq and Faperj (Brazilian agencies), and by TUBITAK (Turkish agency) under the Research Project number 104T148.

References

[1] C. Tsallis, *Introduction to Nonextensive Statistical Mechanics - Approaching a Complex World* (Springer, New York, 2009).
[2] C. Tsallis, *Entropy*, in *Encyclopedia of Complexity and Systems Science*, ed. R.A. Meyers (Springer, Berlin, 2009), 11 volumes [ISBN: 978-0-387-75888-6].
[3] C. Tsallis, *Possible generalization of Boltzmann-Gibbs statistics*, J. Stat. Phys. **52**, 479-487 (1988).
[4] R.J.V. dos Santos, *Generalization of Shannon's theorem for Tsallis entropy*, J. Math. Phys. **38**, 4104 (1997).
[5] S. Abe, *Axioms and uniqueness theorem for Tsallis entropy*, Phys. Lett. A **271**, 74 (2000).
[6] F. Topsoe, *Factorization and escorting in the game-theoretical approach to non-extensive entropy measures*, Physica A **365**, 91-95 (2006).
[7] F. Topsoe, *Towards operational interpretations of generalized entropies*, in the present volume.
[8] A. Ohara, H. Matsuzoe and S. Amari, *A dually flat structure on the space of escort distributions*, in the present volume.
[9] O. Penrose, *Foundations of Statistical Mechanics: A Deductive Treatment* (Pergamon, Oxford, 1970), page 167.
[10] C. Tsallis, M. Gell-Mann and Y. Sato, *Asymptotically scale-invariant occupancy of phase space makes the entropy S_q extensive*, Proc. Natl. Acad. Sc. USA **102**, 15377-15382 (2005).
[11] F. Caruso and C. Tsallis, *Nonadditive entropy reconciles the area law in quantum systems with classical thermodynamics*, Phys. Rev. E **78**, 021101 (2008).
[12] S. Umarov, C. Tsallis and S. Steinberg, *On a q-central limit theorem consistent with nonextensive statistical mechanics*, Milan J. Math. **76**, 307-328 (2008).
[13] S. Umarov, C. Tsallis, M. Gell-Mann and S. Steinberg, *Generalization of symmetric α-stable Lévy distributions for $q \neq 1$*, unpublished (2009).
[14] J.P Bouchaud and A. Georges, *Anomalous diffusion in disordered media: Statistical mechanisms, models and physical applications*, Phys. Rep. **195**, 127 (1990).
[15] A. Rodriguez, V. Schwammle and C. Tsallis, *Strictly and asymptotically scale-invariant probabilistic models of N correlated binary random variables having* q*-Gaussians as $N \to \infty$ limiting distributions*, JSTAT P09006 (2008).
[16] R. Hanel, S. Thurner and C. Tsallis, *Limit distributions of scale-invariant probabilistic models of correlated random variables with the q-Gaussian as an explicit example*, Eur. Phys. J. B (2009), DOI: 10.1140/epjb/e2009-00330.
[17] H.J. Hilhorst and G. Schehr, *A note on q-Gaussians and non-Gaussians in statistical mechanics*, J. Stat. Mech. (2007) P06003.
[18] L.G. Moyano, C. Tsallis and M. Gell-Mann, *Numerical indications of a q-generalised central limit theorem*, Europhys. Lett. **73**, 813-819 (2006).
[19] W.J. Thistleton, J.A. Marsh, K.P. Nelson and C. Tsallis, *q-Gaussian approximants mimic non-extensive statistical-mechanical expectation for many-body probabilistic model with long-range correlations*, Cent. Eur. J. Phys. **7**, 387-394 (2009).
[20] U. Tirnakli, C. Beck and C. Tsallis, *Central limit behavior of deterministic dynamical systems*, Phys. Rev. E **75**, 040106 (2007).
[21] U. Tirnakli, C. Tsallis and C. Beck, *A closer look at time averages of the logistic map at the edge of chaos*, Phys. Rev. E **79**, 056209 (2009).
[22] A. Robledo and M.A. Fuentes, unpublished (2009).
[23] C. Anteneodo and C. Tsallis, *Breakdown of the exponential sensitivity to the initial conditions: Role of the range of the interaction*, Phys. Rev. Lett. **80**, 5313 (1998).
[24] M. Antoni and S. Ruffo, Phys. Rev. E **52**, 2361 (1995).

[25] A. Pluchino, A. Rapisarda and C. Tsallis, *Comment on "Ergodicity and central limit theorem in systems with long-range interactions" by Figueiredo A. et al*, Europhys. Lett. **85**, 60006 (2009).
[26] C. Tsallis, G. Bemski and R.S. Mendes, *Is re-association in folded proteins a case of nonextensivity?*, Phys. Lett. A **257**, 93 (1999).
[27] A. Upadhyaya, J.-P. Rieu, J.A. Glazier and Y. Sawada, *Anomalous diffusion and non-Gaussian velocity distribution of Hydra cells in cellular aggregates*, Physica A **293**, 549 (2001).
[28] A.M. Reynolds, *Can spontaneous cell movements be modelled as Lévy walks?*, Physica A **389**, 273-277 (2010).
[29] K.E. Daniels, C. Beck and E. Bodenschatz, *Defect turbulence and generalized statistical mechanics*, Physica D **193**, 208 (2004).
[30] R. Arevalo, A. Garcimartin and D. Maza, *Anomalous diffusion in silo drainage*, Eur. Phys. J. E **23**, 191-198 (2007)[DOI10.1140/epje/i2006-10174-1].
[31] R. Arevalo, A. Garcimartin and D. Maza, *A non-standard statistical approach to the silo discharge*, in *Complex Systems - New Trends and Expectations*, eds. H.S. Wio, M.A. Rodriguez and L. Pesquera, Eur. Phys. J.-Special Topics **143** (2007) [DOI: 10.1140/epjst/e2007-00087-9].
[32] P. Douglas, S. Bergamini and F. Renzoni, *Tunable Tsallis distributions in dissipative optical lattices*, Phys. Rev. Lett. **96**, 110601 (2006); G.B. Bagci and U. Tirnakli, *Self-organization in dissipative optical lattices*, Chaos **19**, 033113 (2009).
[33] B. Liu and J. Goree, *Superdiffusion and non-Gaussian statistics in a driven-dissipative 2D dusty plasma*, Phys. Rev. Lett. **100**, 055003 (2008).
[34] R.G. DeVoe, *Power-law distributions for a trapped ion interacting with a classical buffer gas*, Phys. Rev. Lett. **102**, 063001 (2009).
[35] L. Borland, *Closed form option pricing formulas based on a non-Gaussian stock price model with statistical feedback*, Phys. Rev. Lett. **89**, 098701 (2002).
[36] L. Borland, *A theory of non-gaussian option pricing*, Quantitative Finance **2**, 415 (2002).
[37] S.M.D. Queiros, *On non-Gaussianity and dependence in financial in time series: A nonextensive approach*, Quant. Finance **5**, 475-487 (2005).
[38] L.F. Burlaga and A.F.-Vinas, *Triangle for the entropic index q of non-extensive statistical mechanics observed by Voyager 1 in the distant heliosphere*, Physica A **356**, 375 (2005).
[39] L.F. Burlaga and N.F. Ness, *Compressible "turbulence" observed in the heliosheath by Voyager 2*, Astrophys. J. **703**, 311-324 (2009).
[40] B. Bakar and U. Tirnakli, *Analysis of self-organized criticality in Ehrenfest's dog-flea model*, Phys. Rev. E **79**, 040103(R) (2009).
[41] F. Caruso, A. Pluchino, V. Latora, S. Vinciguerra and A. Rapisarda, *Analysis of self-organized criticality in the Olami-Feder-Christensen model and in real earthquakes*, Phys. Rev. E **75**, 055101(R)(2007).
[42] L.G. Moyano and C. Anteneodo, *Diffusive anomalies in a long-range Hamiltonian system*, Phys. Rev. E **74**, 021118 (2006).
[43] J.C. Carvalho, R. Silva, J.D. do Nascimento and J.R. de Medeiros, *Power law statistics and stellar rotational velocities in the Pleiades*, Europhys. Lett. **84**, 59001 (2008).
[44] R.M. Pickup, R. Cywinski, C. Pappas, B. Farago and P. Fouquet, *Generalized spin glass relaxation*, Phys. Rev. Lett. **102**, 097202 (2009).

Journal of Physics: Conference Series **201** (2010) 012002 doi:10.1088/1742-6596/201/1/012002

Stationary distributions of sums of marginally chaotic variables as renormalization group fixed points

Miguel Angel Fuentes[1,2,3] **and A. Robledo**[4]

[1] Santa Fe Institute, 1399 Hyde Park Road, Santa Fe, New Mexico 87501, USA
[2] Centro Atómico Bariloche, Instituto Balseiro and CONICET, 8400 Bariloche, Argentina
[3] Center for Advanced Studies in Ecology and Biodiversity, Facultad de Ciencias Biológicas, Pontificia Universidad Católica de Chile, Casilla 114-D, Santiago CP 6513677, Chile
[4] Instituto de Física, Universidad Nacional Autónoma de México, Apartado Postal 20-364, México 01000 DF, Mexico

E-mail: `fuentesm@santafe.edu`

Abstract. We determine the limit distributions of sums of deterministic chaotic variables in unimodal maps assisted by a novel renormalization group (RG) framework associated to the operation of increment of summands and rescaling. In this framework the difference in control parameter from its value at the transition to chaos is the only relevant variable, the trivial fixed point is the Gaussian distribution and a nontrivial fixed point is a multifractal distribution with features similar to those of the Feigenbaum attractor. The crossover between the two fixed points is discussed and the flow toward the trivial fixed point is seen to consist of a sequence of chaotic band mergers.

1. Introduction

The hegemony of the Central Limit Theorem [1, 2] for sums of deterministic variables generated by a number chaotic mappings have for some time been observed and also mathematically proved [3]. Since the mixing properties of chaotic trajectories yield variables indistinguishable to independent random variables, it is of interest to study nonmixing systems such as mappings at the transition from regular to chaotic behavior. Recent [4]-[7] numerical explorations of time averages of iterates at the period-doubling transition to chaos [8] have been presented and interpreted as possible evidence for a novel type of stationary distribution.

The dynamics toward and at the Feigenbaum attractor is now known in much detail [9, 10], therefore, it appears feasible to analyze also the properties of sums of iterate positions for this classic nonlinear system with the same kind of analytic reasoning and numerical thoroughness. Here we present the results for sums of chronological positions of trajectories associated to quadratic unimodal maps. We consider the case of the sum of positions of trajectories inside the Feigenbaum attractor as well as those within the chaotic 2^n-band attractors obtained when the control parameter is shifted to values larger than that at the transition to chaos. From the information obtained we draw conclusions on the properties of the stationary distributions for these sums of variables. Our results, that reveal a multifractal stationary distribution that mirrors the features of the Feigenbaum attractor, can be easily extended to other critical attractor universality classes and other routes to chaos.

The overall picture we obtain is effectively described within the framework of the renormalization group (RG) approach for systems with scale invariant states or attractors. Firstly, the RG transformation for the distribution of a sum of variables is naturally given by the change due to the increment of summands followed by a suitable restoring operation. Second, the limit distributions can be identified as fixed points reached according to whether the acting relevant variables are set to zero or not. Lastly, the universality class of the non-trivial fixed-point distribution can be assessed in terms of the existing set of irrelevant variables.

As it is well known [8] a few decades ago the RG approach was successfully applied to the period-doubling route to chaos displayed by unimodal maps. In that case the RG transformation is functional composition of the mapping and its effect re-enacts the growth of the period doubling cascade. In our case the RG transformation is the growth and adjustment of the sum of positions and its effect is instead to go over again the merging of bands in the chaotic region.

Specifically, we consider the Feigenbaum map $g(x)$, obtained from the fixed point equation $g(x) = \alpha g(g(x/\alpha))$ with $g(0) = 1$ and $g'(0) = 0$, and where $\alpha = -2.50290...$ is one of Feigenbaum's universal constants [8]. For expediency we shall from now on denote the absolute value $|\alpha|$ by α. Numerically, the properties of $g(x)$ can be conveniently obtained from the logistic map $f_{\mu,2}(x) = 1 - \mu x^2$, $-1 \leq x \leq 1$, with $\mu = \mu_\infty = 1.401155189092...$ The dynamics associated to the Feigenbaum map is determined by its multifractal attractor. For a recent detailed description of these properties see [9, 10]. For values of $\mu > \mu_\infty$ we employ a well-known scaling relation supported by numerical results.

Initially we present properties of the sum of the absolute values $|x_t|$ of positions $x_t = f_{\mu_\infty,2}(x_{t-1})$, $t = 1, 2, 3, ...$, as a function of total time N visited by the trajectory with initial position $x_0 = 0$, and obtain a patterned linear growth with N. We analyze this intricate fluctuating pattern, confined within a band of finite width, by eliminating the overall linear increment and find that the resulting stationary arrangement exhibits features inherited from the multifractal structure of the attractor. We derive an analytical expression for the sum that corroborates the numerical results and provide an understanding of its properties. Next, we consider the straight sum of x_t, where the signs taken by positions lessen the growth of its value as N increases and the results are consistently similar to those for the sum of $|x_t|$, i.e. linear growth of a fixed-width band within which the sum displays a fluctuating arrangement. Further details for the sum of x_t are not included because of repetitiveness. Then, we show numerical results for the sum of iterated positions obtained when the control parameter is shifted into the region of chaotic bands. In all of these cases the distributions evolve after a characteristic crossover towards a Gaussian form. Finally, we rationalize our results in terms of an RG framework in which the action of the Central Limit Theorem plays a fundamental role.

The starting point of our study is evaluation of

$$y_\mu(N) \equiv \sum_{t=1}^{N} |x_t| , \tag{1}$$

with $\mu = \mu_\infty$ and with $x_0 = 0$. Fig. 1A shows the result, where it can be observed that the values recorded, besides a repeating fluctuating pattern within a narrow band, increase linearly on the whole. The measured slope of the linear growth is $c = 0.56245...$ Fig. 1B shows an enlargement of the band, where some detail of the complex pattern of values of $y_{\mu_\infty}(N)$ is observed. A stationary view of the mentioned pattern is shown in Fig. 1C, where we plot

$$y'_{\mu_\infty}(N) \equiv \sum_{t=1}^{N} (|x_t| - c) , \tag{2}$$

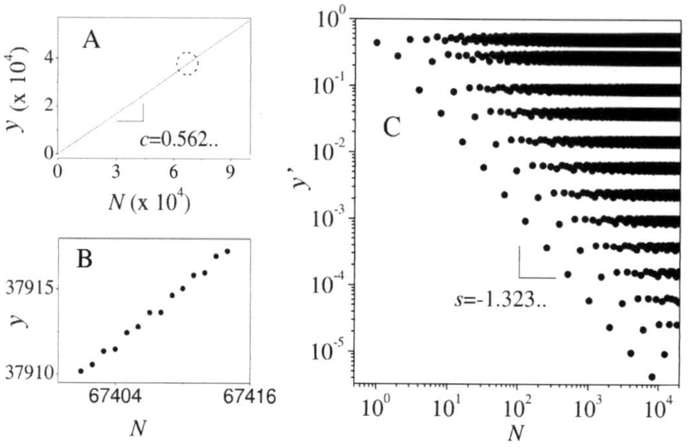

Figure 1. A) Sum of absolute values of visited points x_t, $t = 0, ..., N$, of the Feigenbaum's attractor with initial condition $x_0 = 0$. B) A closer look of the path of the sum (see dotted circle in A), for values of N around 67410. C) Centered sum $y'(N)$ in logarithmic scales. See text.

in logarithmic scales. There, we observe that the values of $y'_{\mu_\infty}(N)$ fall within horizontal bands interspersed by gaps, revealing a fractal or multifractal set layout. The top (zeroth) band contains y'_{μ_∞} for all the odd values of N, the 1st band next to the top band contains y'_{μ_∞} for the even values of N of the form $N = 2+4m$, $m = 0, 1, 2, ...$ The 2nd band next to the top band contains $y'_{\mu_\infty}(N)$ for $N = 2^2 + 2^3 m$, $m = 0, 1, 2, ...$, and so on. In general, the k-th band next to the top band contains $y'_{\mu_\infty}(2^k + 2^{k+1}m)$, $m = 0, 1, 2, ...$ Another important feature in this figure is that the $y'_{\mu_\infty}(N)$ for subsequences of N each of the form $N = (2l + 1)2^k$, $k = 0, 1, 2, ...$, with l fixed at a given value of $l = 0, 1, 2, ...$, appear aligned with a uniform slope $s = -1.323...$ The parallel lines formed by these subsequences imply the power law $y'_{\mu_\infty}(N) \sim N^s$ for N belonging to such a subsequence.

It is known [9, 11] that these two characteristics of $y'_{\mu_\infty}(N)$ are also present in the layout of the absolute value of the individual positions $|x_t|$, $t = 1, 2, 3, ...$ of the trajectory initiated at $x_0 = 0$; and this layout corresponds to the multifractal geometric configuration of the points of the Feigenbaum's attractor, see Fig. 1 in [11]. In this case, the horizontal bands of positions separated by equally-sized gaps are related to the period-doubling 'diameters' [8] set construction of the multifractal [10]. The identical slope shown in the logarithmic scales by all the position subsequences $|x_t|$, $t = (2l + 1)2^k$, $k = 0, 1, 2, ...$, each formed by a fixed value of $l = 0, 1, 2, ...$, implies the power law $|x_t| \sim t^s$, $s = -\ln\alpha/\ln 2 = -1.3236...$, as the $|x_t|$ can be expressed as $|x_t| \simeq |x_{2l+1}|\ \alpha^{-k}, t = (2l + 1)2^k$, $k = 0, 1, 2, ...$, or, equivalently, $|x_t| \sim t^s$. Notice that the index k also labels the order of the bands from top to bottom. The power law behavior involving the universal constant α of the subsequence positions reflect the approach of points in the attractor toward its most sparse region at $x = 0$ from its most compact region, as the positions at odd times $|x_{2l+1}| = x_{2l+1}$, those in the top band, correspond to the densest region of the set.

Having uncovered the through manifestation of the multifractal structure of the attractor into the sum $y'_{\mu_\infty}(N)$ we proceed to derive this property and corroborate the numerical evidence. Consider Eq. (1) with $N = 2^k$, $k = 0, 1, 2, ...$, the special case $l = 0$ in the discussion above. Then the numbers of terms $|x_t|$ per band in $y_{\mu_\infty}(2^k)$ are: 2^{k-1} in the top band ($j = 0$), 2^{k-2}

in the next band ($j = 1$),..., 2^0 in the ($k-1$)-th band, plus an additional position in the k-th band. If we introduce the average of the positions on the top band

$$\langle a \rangle \equiv 2^{-(k-1)} \sum_{j=0}^{2^{k-1}} x_{2j+1}, \tag{3}$$

the sum $y_{\mu_\infty}(2^k)$ can be written as

$$y_{\mu_\infty}(2^k) = \langle a \rangle \, 2^{k-1} \sum_{j=0}^{k-2} (2\alpha)^{-j} + \alpha^{-(k-1)} + \alpha^{-k}. \tag{4}$$

Doing the geometric sum above and expressing the result as $y_{\mu_\infty}(2^k) = c2^k + d\alpha^{-k}$, we have

$$c = \frac{\langle a \rangle \, \alpha}{2\alpha - 1}, \quad d = \left(1 - \frac{\langle a \rangle \, 2\alpha}{2\alpha - 1}\right) \alpha + 1 \,. \tag{5}$$

Evaluation of Eq. (3) yields to $\langle a \rangle = 0.8999...$, and from this we obtain $c = 0.56227...$ and $d = 0.68826...$ We therefore find that the value of the slope c in Fig. 1A is properly reproduced by our calculation. Also, since $\ln \left[y_{\mu_\infty}(2^k) - c2^k \right] = \ln d - k \ln \alpha$, or, equivalently, $\ln y'_{\mu_\infty}(N) = \ln d - N \ln \alpha / \ln 2$, $N = 2^k$, $k = 0, 1, 2, ...$, we corroborate that the value of the slope s in inset of Fig. 1C is indeed given by $s = -\ln \alpha / \ln 2 = 1.3236...$ (We have made use of the identity $\alpha^{-k} = N^{-\ln \alpha / \ln 2}$, $N = 2^k$, $k = 0, 1, 2, ...$).

We note that the sum of x_t from $t = 0$ to $N = 2^k$, i.e. considering the signs taken by positions, can be immediately obtained from the above by replacing α^{-j} by $(-1)^j \alpha^{-j}$ as the x_t of different signs of the trajectory starting at $x_0 = 0$ fall into separate alternating bands (described above and shown in Fig. 1 of [11]). In short, $x_t \simeq (-1)^j x_{2l+1} \, \alpha^{-j}$, $t = (2l+1)2^k$, $k = 0, 1, 2, ...$ As stated, our numerical and analytical results are in agreement also in this case.

We turn now to study the sum of positions of trajectories when $\Delta\mu \equiv \mu - \mu_\infty > 0$. We recall that in this case the attractors are made up of 2^K, $K = 1, 2, 3, ...$, bands and that their trajectories consist of an interband periodic motion of period 2^K and an intraband chaotic motion. We evaluated numerically the sums $y_\mu(N)$ for an ensemble of initial conditions x_0 uniformly distributed only within the chaotic bands, for different values of $\Delta\mu$; $y'_\mu(N)$ was then obtained similarly to Eq. (2) by substracting the average $\langle y_\mu(N) \rangle_{x_0}$ and rescaling with a factor $N^{-1/2}$. The panels in Fig. 2 show the evolution of the distributions for increasing number of summands N for a value of $\Delta\mu$ (chosen for visual clarity) when the attractor consists of 2^3 chaotic bands. Initially the distributions are multimodal with disconnected domains, but as N increases we observe merging of bands and development of a single-domain bell-shaped distribution that as $N \longrightarrow \infty$ converges in all cases to a Gaussian distribution.

These numerical results can be understood as follows. We recollect [8] that the relationship between the number 2^K, $K \gg 1$, of bands of a chaotic attractor and the control parameter distance $\Delta\mu$ at which it is located is given by $2^K \sim \Delta\mu^{-\kappa}$, $\kappa = \ln 2 / \ln \delta_F$, where $\delta_F = 0.46692...$ is the universal constant that measures both the rate of convergence of the values of μ at period doublings or at band splittings to μ_∞. For $\Delta\mu$ small and fixed, the sum of sequential positions of the trajectory initiated at $x_0 = 0$, Eq. (1), exhibits two different growth regimes as the total time N increases. In the first one, when $N \ll 2^K$, the difference in value $\delta x_t \equiv x_t(\mu) - x_t(\mu_\infty)$ between the positions at time t for μ and μ_∞ do not affect qualitatively the multifractal structure of the sum at μ_∞, nor its associated distribution. This is because the fine structure of the Feigenbaum attractor is not suppressed by the fluctuations δx_t, as these contribute to the sum individually during the first cycle of the interband periodic motion. The discrete multi-scale

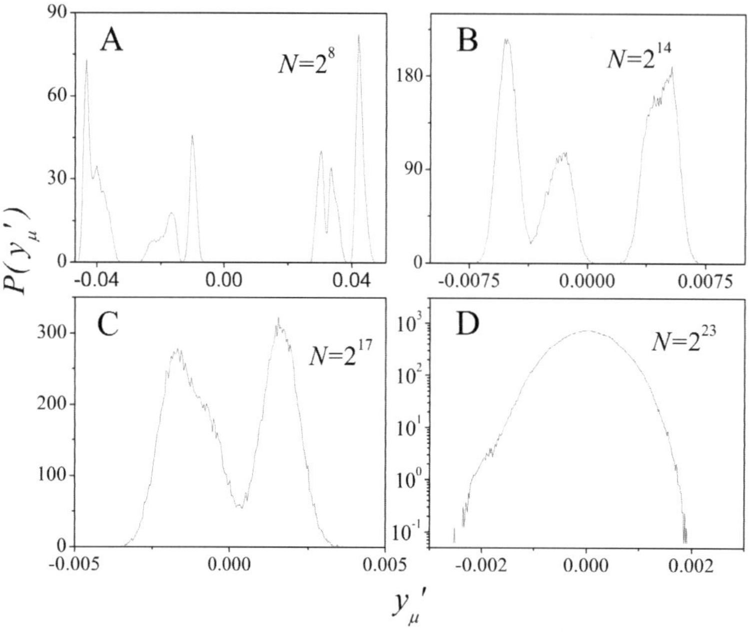

Figure 2. Distributions for the sums of $|x_t|$, $t = 0, ..., N$, of an ensemble of trajectories with initial conditions within the 2^3-band attractor at $\Delta\mu = 0.0028448109$. The number of summands N are indicated in each panel. See text.

nature of the distribution for μ_∞ is preserved when the interband motion governs the sum $y_\mu(N)$. In the second regime, when $N \gg 2^K$, the situation is opposite, after many interband cycles the fluctuations δx_t add up in the sum and progressively wipe up the fine structure of the Feigenbaum attractor, leading to merging of bands and to the dominance of the fluctuating intraband motion. Ultimately, as $N \longrightarrow \infty$ the evolution of the distribution is similar to the action of the Central Limit Theorem and leads to a Gaussian stationary end result. It is also evident that as $\Delta\mu$ increases the first regime is shortened at the expense of the second, whereas when $\Delta\mu \longrightarrow 0$ the converse is the case. Therefore there exists an unambiguous $\Delta\mu$-dependent crossover behavior between the two radically different types of stationary distributions. This crossover is set out when the δx_t fluctuations begin removing the band structure in $y'_\mu(N)$ when $\Delta\mu$ is small and ends when these fluctuations have broadened and merged all the chaotic bands and $y'_\mu(N)$ forms a single continuous interval. When $\mu = \mu_\infty$ this process never takes place.

We are in a position now to put together the numerical and analytical information presented above into the general framework of the RG approach. As known, this method was designed to characterize families of systems containing amongst their many individual states (or in this case attractors) a few exceptional ones with scale invariant properties and common to all systems in the family. We recall [12] that in the language of a minimal RG scheme there are two fixed points, each of which can be reached by the repeated application of a suitable transformation of the system's main defining property. One of the fixed points, is termed trivial and is reached via the RG transformation for almost all initial settings. i.e. for all systems in the family when at least one of a small set of variables, named relevant variables, is nonzero. To reach the other fixed point, termed nontrivial, it is necessary that the relevant variables are all set to zero, and this implies a severely restricted set of initial settings that ensure such critical RG paths. The nontrivial fixed point embodies the scale invariant properties of the exceptional state that occurs in each system in the family and defines a universality class, while the differences amongst the individual systems are distinguished through a large set of so-called irrelevant variables. The variables in the latter set gradually vanish as the RG transformation is applied to a system that evolves toward the nontrivial fixed point. Further, when any system in the family is given a nonzero but sufficiently small value to (one or more of) the relevant variables, the RG transformation converts behavior similar to that of the nontrivial fixed point into that resembling the trivial fixed point through a well-defined crossover phenomenon. The recognition of the RG framework in the properties of the sums of positions of trajectories in unimodal maps and their associated distributions is straightforward. It can be concluded right away that in this problem (as defined here) there is only one relevant variable, the control parameter difference $\Delta\mu$. There is an infinite number of irrelevant variables, those that specify the differences between all possible unimodal maps (with quadratic maximum) and the Feigenbaum map $g(x)$. There are two fixed-point distributions, the trivial continuum-space Gaussian distribution and the nontrivial discrete-space multifractal distribution (as observed in Fig. 1C). As explained above, there is a distinct crossover link between the two fixed-point distributions. The RG transformation consists of the increment of one or more summands in the sum (1) followed by centering like in Eq. (2). The effect of the transformation in the distribution of the sum is then recorded. For sums of independent variables the transformation is equivalent to the convolution of distributions. Our results correspond to the dynamics inside the atractors, however, if the interest lies in considering only the stationary distribution of sums that do not contain the transient behavior of trajectories in their way to the attractor [4] our results are expected to give the correct answers for this case.

In summary, we have found that the stationary distribution of the sum of iterate positions within the Feigenbaum attractor has a multifractal structure stamped by that of the initial multifractal set, while that involving sums of positions within the attractors composed of 2^K chaotic bands is the Gaussian distribution. We have also shown that the entire problem can be couched in the language of the RG formalism in a way that makes clear the identification of the existing stationary distributions and the manner in which they are reached. These basic features suggest a degree of universality, limited to the critical attractor under consideration, in the properties of sums of deterministic variables at the transitions to chaos. Namely, the sums of positions of memory-retaining trajectories evolving under a vanishing Lyapunov exponent appear to preserve the particular features of the multifractal critical attractor under examination. Thus we expect that varying the degree of nonlinearity of a unimodal map would affect the scaling properties of time averages of trajectory positions at the period doubling transition to chaos, or alternatively, that the consideration of a different route to chaos, such as the quasiperiodic route, would lead to different scaling properties of comparable time averages. For instance, the known dependence of the universal constant α on the degree of nonlinearity z of a unimodal map would show as a z-dependent exponent $s = \ln\alpha/\ln 2$ that controls the scale invariant property of the sum of trajectory positions with $x_0 = 0$ (shown in Fig. 1C).

We analyzed the nature and the conditions under which a stationary distribution with universal properties (in the Renormalization Group sense) occurs for sums of deterministic variables at the transition between regular and chaotic behavior, such as those studied here for dynamics at zero Lyapunov exponent. The nonexistence of fluctuations in such critical attractors implies a distribution of the sums of these variables strictly defined on a discrete multifractal set and therefore different from known (Gaussian or otherwise) continuum-space limit distributions for real number random variables.

Acknowledgments
We appreciate partial financial support by DGAPA-UNAM and CONACYT (Mexican agencies). AR is grateful for hospitality received at the SFI.

References

[1] van Kampen N G 1981 *Stochastic Processes in Physics and Chemistry* (North-Holland, Amsterdam)
[2] Khinchin A Y 1949 *Mathematical Foundations of Statistical Mechanics* (Dover, New York)
[3] Mackey M C and Tyran-Kaminska M 2006 *Phys. Rep.* **422** 167
[4] Tirnakli U, Beck C and Tsallis C 2007 *Phys. Rev. E* **75** 040106(R)
[5] Tirnakli U, Beck C and Tsallis C 2009 *Phys. Rev. E* **79** 056209
[6] Tirnakli U, Tsallis C and Beck C (arXiv:0906.1262) [cond-mat.stat-mech]
[7] Grassberger P 2009 *Phys. Rev. E* **79** 057201
[8] Schuster H G 1988 *Deterministic Chaos. An Introduction* (2nd Revised Edition, VCH Publishers, Weinheim)
[9] Mayoral E and Robledo A 2005 *Phys. Rev. E* **72** 026209
[10] Robledo A and Moyano L G 2008 *Phys. Rev. E* **77** 036213
[11] Robledo A 2006 *Physica A* **370** 449
[12] Fisher M E 1998 *Rev. Mod. Phys.* **70** 653

Journal of Physics: Conference Series **201** (2010) 012003 doi:10.1088/1742-6596/201/1/012003

The q-exponential family in statistical physics

Jan Naudts

Universiteit Antwerpen, Groenenborgerlaan 1717, 2020 Antwerpen, Belgium

E-mail: jan.naudts@ua.ac.be

Abstract. The Boltzmann-Gibbs probability distribution, seen as a statistical model, belongs to the exponential family. Recently, the latter concept has been generalized. The q-exponential family has been shown to be relevant for the statistical description of small isolated systems. Two main applications are reviewed: 1. The distribution of the momentum of a single particle is a q-Gaussian, the distribution of its velocity is a deformed Maxwellian; 2. The configurational density distribution belongs to the q-exponential family.

The definition of the temperature of small isolated systems is discussed. It depends on defining the thermodynamic entropy of a microcanonical ensemble in a suitable manner. The simple example of non-interacting harmonic oscillators shows that Rényi's entropy functional leads to acceptable results.

1. Introduction

In statistics, a model is a probability distribution which depends on a number of parameters. The probability distributions of statistical physics usually depend on parameters and can therefore be seen as statistical models. Typical parameters are the total energy U in the microcanonical ensemble, or the inverse temperature β in the canonical ensemble. This parameter dependence is important in order to understand why certain models belong to the exponential family and others do not. In particular, all models described by a Boltzmann-Gibbs distribution belong to the exponential family because they have the right dependence on the inverse temperature β.

Recently, the notion of the exponential family has been generalized by the present author in a series of papers [1, 2, 3, 4, 5, 6]. The same definition of generalized exponential family has been introduced in the mathematics literature [1, 7, 8, 9, 5]. This class of models was also derived using the maximum entropy principle in [10, 11] and in the context of game theory [12, 13]. Here we concentrate on the more specific notion of the q-exponential family and review the evidence that this family appears in a natural manner in the context of the microcanonical description of a classical gas.

The notion of q-exponential family is connected with Amari's α-family [14], studied in the context of information geometry. The geometric approach is very appealing also in the context of statistical physics. See for instance [15, 16]. However, this topic will not be discussed in the present paper.

Note that in many of the formulas given below Boltzmann's constant k_B is set equal to 1, which means that temperatures are measured in units of energy.

Journal of Physics: Conference Series **201** (2010) 012003 doi:10.1088/1742-6596/201/1/012003

2. The Boltzmann-Gibbs distribution

The Boltzmann-Gibbs (BG) distribution is given by

$$f_\beta(x) = \frac{c(x)}{Z(\beta)} e^{-\beta H(x)}, \qquad x \text{ in phase space} \tag{1}$$

The *Hamiltonian* $H(x)$ is a given function. It determines the *physical model*. β is a parameter, corresponding with the inverse temperature of thermodynamics. The function $c(x)$ quite often is just a constant. $Z(\beta)$ is called the *partition sum*. It normalizes the probability distribution:

$$Z(\beta) = \int \mathrm{d}x\, c(x) e^{-\beta H(x)}. \tag{2}$$

$\Phi(\beta) \equiv \ln Z(\beta)$ corresponds in thermodynamics with Massieu's function. It satisfies the identity

$$\frac{\mathrm{d}\Phi}{\mathrm{d}\beta} = -U \tag{3}$$

with U the *energy*, this is, the expected value of the Hamiltonian $H(x)$

$$U = \mathbb{E}_\beta H = \int \mathrm{d}x\, f_\beta(x) H(x). \tag{4}$$

The second derivative of Massieu's function satisfies

$$\frac{\mathrm{d}^2\Phi}{\mathrm{d}\beta^2} = \mathbb{E}_\beta (H-U)^2 \geq 0. \tag{5}$$

Therefore, $\Phi(\beta)$ is a convex function, as expected from thermodynamics. Indeed, in thermodynamics Massieu's function is by definition [17] the Legendre transform of the *thermodynamic entropy* $S(U)$, which is a function of the energy U. Within statistical physics the thermodynamic entropy can be defined by means of an *entropy functional*, $I(f)$, which is a function over the space of all probability distributions f. The most common entropy functional is that of Boltzmann-Gibbs-Shannon (BGS)

$$I(f) = -\int \mathrm{d}x\, f(x) \ln \frac{f(x)}{c(x)}. \tag{6}$$

It is well-known that $I(f)$ takes its maximal value at $f = f_\beta$ within the set of all f satisfying $\mathbb{E}_f H = U$. This maximal value of $I(f)$ is then identified with the thermodynamic entropy $S(U)$. It satisfies

$$\begin{aligned} S(U) &\equiv \max_f \{I(f) : \mathbb{E}_f H = U\} = I(f_\beta) \\ &= \int \mathrm{d}x\, f_\beta(x) \left(\ln Z(\beta) + \beta H(x)\right) \\ &= \ln Z(\beta) + \beta U. \end{aligned} \tag{7}$$

Comparison with

$$\Phi(\beta) = \max_U \{S(U) - \beta U\} \tag{8}$$

then yields $\Phi(\beta) = \ln Z(\beta)$. From (7, 8) follows the inverse transformation

$$S(U) = \inf_\beta \{\Phi(\beta) + \beta U\} \tag{9}$$

and an identity, which is the dual expression of (3),

$$\beta = \frac{\mathrm{d}S}{\mathrm{d}U}. \tag{10}$$

The latter relation is often used as the definition of the inverse temperature. Another consequence of (9) is that $S(U)$ is a concave function.

3. The exponential family

A model of classical (statistical) mechanics is determined by a Hamiltonian $H(x)$. On the other hand, a *statistical model* is defined as a parameterized probability distribution function (pdf). It is now immediately clear that the Boltzmann-Gibbs distribution (1) defines a statistical model with one parameter β. Moreover, it belongs to a very special class of models called *the exponential family*. In fact, a model belongs to the exponential family with parameters $\theta = (\theta_1, \theta_2, \cdots, \theta_n)$ when its pdf $f_\theta(x)$ can be written into the canonical form

$$f_\theta(x) = c(x) \exp\left(\sum_{j=1}^{n} \theta_j K_j(x) - \alpha(\theta) \right), \tag{11}$$

where $c(x)$, $K_j(\theta)$, and $\alpha(\theta)$ are known functions.

Models belonging to the exponential family share a number of nice properties. For instance, they satisfy the identities, which generalize (3),

$$\frac{\partial \alpha}{\partial \theta_j} = \mathbb{E}_\theta K_j = \int \mathrm{d}x\, f_\theta(x) K_j(x). \tag{12}$$

Note that the functions $K_j(x)$ may not depend on the parameters θ. In statistical physics one uses sometimes effective Hamiltonians which depend on the temperature β^{-1}. The corresponding statistical models may exhibit thermodynamic instabilities which cannot occur in models belonging to the exponential family.

4. Deformed logarithmic and exponential functions

The Boltzmann-Gibbs distribution is so dominant in statistical physics that one can get the impression that this is the only relevant statistical model. However there exists a much larger class of models, sharing most of the nice properties of the exponential family. Some of these more general models appear in a natural manner in statistical physics, as will be shown further on. One way to get access to this larger class is by replacing the exponential function in the definition (11) by some other function. This can be done in a fairly general way [1, 11]. Here, a one-parameter deformation is considered, which is the basis for the generalization known as *non-extensive statistical mechanics* [18].

Fix a real number $q \neq 1$. The q-deformed logarithm is defined by [19, 20]

$$\ln_q(u) = \frac{1}{1-q}\left(u^{1-q} - 1\right), \qquad u > 0. \tag{13}$$

The basic properties of the q-deformed logarithmic function are that $\ln_q(1) = 0$ and that it is a strictly increasing function. Indeed, its first derivative is always positive

$$\frac{\mathrm{d}}{\mathrm{d}u} \ln_q(u) = \frac{1}{u^q} > 0 \tag{14}$$

For $q > 0$ the q-logarithm is a concave function. In the limit $q = 1$ it reduces to the natural logarithm $\ln u$.

The inverse function is the deformed exponential function

$$\exp_q(u) = [1 + (1-q)u]_+^{1/(1-q)}. \tag{15}$$

The notation $[u]_+ = \max\{0, u\}$ is used. One has $0 \leq \exp_q(u) \leq +\infty$ for all u. For $q \neq 1$ the range of $\ln_q(u)$ is not the full line. By putting $\exp_q(u) = 0$ when u is below the range of $\ln_q(u)$,

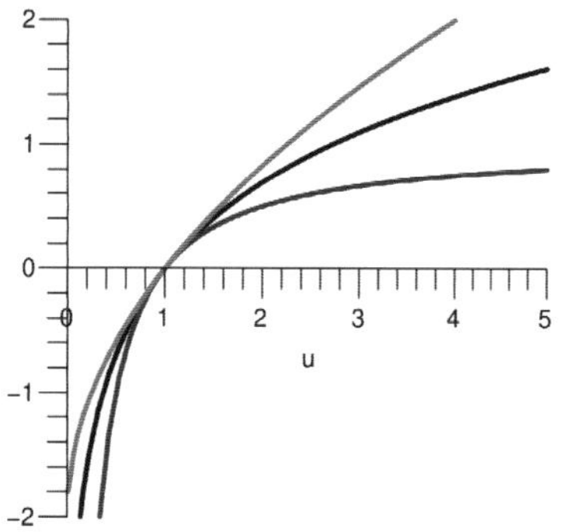

Figure 1. The q-deformed logarithm $\ln_q(u)$ for q-values 0.5 (red), 1 (black), and 2 (blue).

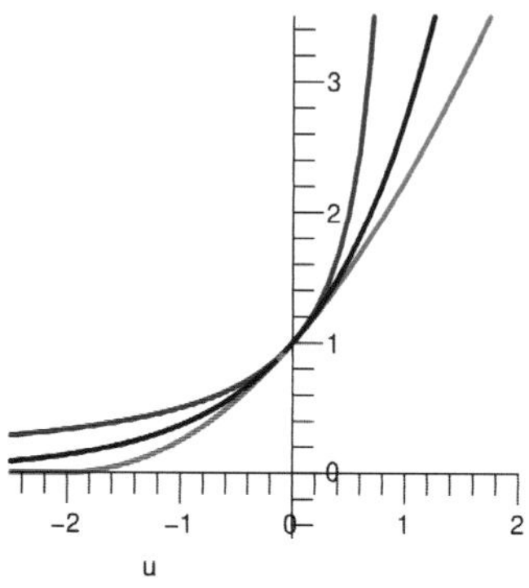

Figure 2. The q-deformed exponential $\exp_q(u)$ for q-values 0.5 (red),1 (black), and 2 (blue).

and equal to $+\infty$ when it is above, $\exp_q(u)$ is an increasing function of u, defined for all values of u. With this extended definition one has

$$\exp_q(\ln_q(u)) = u \qquad \text{for all } u > 0. \tag{16}$$

However, the relation $\ln_q(\exp_q(u)) = u$ is only defined when $\exp_q(u)$ is neither 0 or $+\infty$.

The basic properties of the q-deformed exponential are that $\exp_q(0) = 1$ and that it is an increasing function. Indeed, one has

$$\frac{\mathrm{d}}{\mathrm{d}u}\exp_q(u) = \left[\exp_q(u)\right]^q \geq 0. \tag{17}$$

For $q > 0$ it is a convex function.

5. The q-exponential family

The deformed logarithmic and exponential functions can be used in more than one way to develop a generalized formalism of statistical physics. The original approach of non-extensive statistical physics [21] was to first generalize the Boltzmann-Gibbs-Shannon entropy functional (6) by deforming the logarithm contained in it. Next one studies pdfs which maximize this entropy functional under given constraints. However, it turns out to be advantageous to generalize immediately the notion of exponential family by deforming the exponential function appearing in (11). This leads to the definition that the one-parameter model $f_\theta(x)$ belongs to the *q-exponential family* if there exist functions $c(x), \alpha(\theta), H(x)$ such that one can write

$$f_\theta(x) = c(x)\exp_q\left(-\alpha(\theta) - \theta H(x)\right). \tag{18}$$

In the limit $q = 1$ this reduces to the standard definition of the exponential family with $\theta = \beta$. Note that we take into account the possibility that the parameter θ does not necessarily coincide with the inverse temperature β as it is defined by the thermodynamic relation (10).

Introduce the q-deformed entropy functional

$$I_q(f) \quad = \quad -\int \mathrm{d}x\, f(x) \ln_q \frac{f(x)}{c(x)}$$

$$= \frac{1}{1-q}\left(1 - \int \mathrm{d}x\, c(x) \left(\frac{f(x)}{c(x)}\right)^{2-q}\right). \tag{19}$$

In the limit $q = 1$ it reduces to the BGS-entropy functional (6). Note that this definition is almost the entropy functional originally introduced by Tsallis [21], but differs from it by the fact that the parameter q has been substituted by $2 - q$.

Assume now $0 < q < 2$. Then one can show that the pdf $f = f_\theta(x)$ maximizes the quantity

$$\frac{1}{2-q} I_q(f) - \theta\,\mathbb{E}_f H. \tag{20}$$

In the $q = 1$-case this result is known as *the variational principle*. A proof of this statement involves the introduction of an appropriate Bregman divergence — see [1, 5]. Note that the variational principle implies *the maximum entropy principle*: $I_q(f_\theta) \geq I_q(f)$ for all f which satisfy $\mathbb{E}_f H = \mathbb{E}_\theta H$ (using the notation $\mathbb{E}_\theta \equiv \mathbb{E}_{f_\theta}$). A well-known problem is now that f_θ also maximizes $\xi(I_q(f))$, where $\xi(u)$ is an arbitrary monotonically increasing function. Therefore a meaningful *ansatz* is to assume that the thermodynamic entropy $S(U)$ is given by

$$S(U) = \xi\left(I_q(f_\theta)\right) = \xi\left(\alpha(\theta) + \theta\,\mathbb{E}_\theta H\right). \tag{21}$$

The thermodynamic expression for the inverse temperature β can then be calculated using (10). It is clear that the resulting relation between energy U and temperature β^{-1} will depend on the choice of the monotonic function $\xi(u)$. Physical arguments are needed to decide which choice is the right one.

6. q-Gaussians

The q-Gaussian distribution in one variable is given by (see for instance [22, 23, 24])

$$f(x) = \frac{1}{c_q \sigma} \exp_q(-x^2/\sigma^2), \tag{22}$$

with

$$\begin{aligned} c_q = \int_{-\infty}^{\infty} \mathrm{d}x\, \exp_q(-x^2) &= \sqrt{\frac{\pi}{q-1}}\frac{\Gamma\left(-\frac{1}{2}+\frac{1}{q-1}\right)}{\Gamma\left(\frac{1}{q-1}\right)} \quad \text{if} \quad 1 < q < 3, \\ &= \sqrt{\frac{\pi}{1-q}}\frac{\Gamma\left(1+\frac{1}{1-q}\right)}{\Gamma\left(\frac{3}{2}+\frac{1}{1-q}\right)} \quad \text{if} \quad q < 1. \end{aligned} \tag{23}$$

It can be brought into the form (18) with $c(x) = 1/c_q$, $H(x) = x^2$, $\theta = \sigma^{3-q}$, and

$$\alpha(\theta) = \frac{\sigma^{q-1} - 1}{q-1} = \ln_{2-q}(\sigma). \tag{24}$$

The $q = 1$-case reproduces the conventional Gauss distribution. For $q < 1$ the distribution vanishes outside an interval. Take for instance $q = 1/2$. Then (22) becomes

$$f(x) = \frac{15\sqrt{2}}{32\sigma}\left[1 - \frac{x^2}{\sigma^2}\right]_+^2. \tag{25}$$

This distribution vanishes outside the interval $[-\sigma, \sigma]$. In the range $1 \leq q < 3$ the q-Gaussian is strictly positive on the whole line and decays with a power law in $|x|$ instead of exponentially. For $q = 2$ one obtains

$$f(x) = \frac{1}{\pi} \frac{\sigma}{x^2 + \sigma^2}. \tag{26}$$

This is the Cauchy distribution. The function (26) is also called a Lorentzian and is often used in physics to fit the shape of spectral lines. For $q \geq 3$ the distribution cannot be normalized because

$$f(x) \sim \frac{1}{|x|^{2/(q-1)}} \text{ as } |x| \to \infty \tag{27}$$

is not integrable anymore.

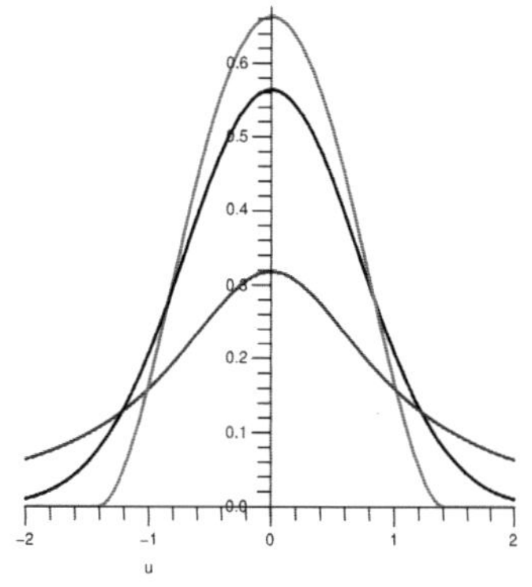

Figure 3. q-Gaussians for $\sigma = 1$ and for q-values 0.5 (red), 1 (black), and 2 (blue).

7. A q-deformed Maxwell distribution

The most straightforward way to produce an important example of q-Gaussians in statistical physics is by considering marginals of the uniform distribution on a hypersphere – see [25]. The Hamiltonian of an N-particle classical ideal gas is given by

$$H(p) = \frac{1}{2m} \sum_{j=1}^{N} |\mathbf{p}_j|^2, \tag{28}$$

where m is the mass of the particles and $\mathbf{p}_j$ is the momentum of the j-th particle. Given the value E for the total energy, the phase space consists of all points of the $3N$-dimensional sphere with radius $\sqrt{2mE}$. Assume equal probability of all points of phase space. Let $B_n(r)$ denote the volume of a sphere with radius r in dimension n. The probability distribution for the momentum of a single particle becomes

$$\begin{aligned} f(\mathbf{p}_1) &= \frac{1}{(2m)^{3N/2} B'_{3N}(E)} \int \mathrm{d}^3\mathbf{p}_2 \cdots \mathrm{d}^3\mathbf{p}_N \, \delta(E - H(p)) \\ &= \frac{1}{(2m)^{3/2}} \frac{B'_{3(N-1)}(E - |\mathbf{p}_1|^2/2m)}{B'_{3N}(E)} \\ &= \frac{1}{c_q \sigma} \exp_q(-|\mathbf{p}_1|^2/2\sigma^2) \end{aligned} \tag{29}$$

with

$$q = \frac{3N-6}{3N-4} \quad \text{and} \quad \sigma^2 = \frac{2mE}{3N-4}. \tag{30}$$

This is a q-Gaussian with $q < 1$. The appearance of a cutoff can be easily understood. Arbitrary large momenta are not possible because of the obvious upperbound $|\mathbf{p}_1|^2 \leq 2mE$. The probability distribution of the scalar velocity $v = |\mathbf{p}_1|/m$ becomes

$$f_q(v) = \frac{4\pi m^3}{c_q \sigma} v^2 \exp_q\left(-\frac{1}{2\sigma^2} m v^2\right). \tag{31}$$

Only in the limit of large systems it converges to the Maxwell distribution. One concludes that the Maxwell distribution is an approximation and is only valid in the limit $N \to \infty$.

8. Configurational density of classical gases

The configurational density of a gas of classical particles always belongs to the q-exponential family [26]. The idea behind the calculation is the same as that of the previous section. This kind of non-extensivity of a classical gas was observed already quite some time ago [27, 28]. It becomes a strong statement in the language of q-exponential families.

A classical gas of N interacting particles is described by the Hamiltonian

$$H(\mathbf{q}, \mathbf{p}) = \frac{1}{2m} \sum_{j=1}^{N} |\mathbf{p}_j|^2 + \mathcal{V}(\mathbf{q}), \tag{32}$$

where $\mathcal{V}(\mathbf{q})$ is the potential energy due to interaction among the particles and between the particles and the walls of the system. The microcanonical ensemble is then by definition the following singular probability distribution

$$f_U(\mathbf{q}, \mathbf{p}) = \frac{1}{\omega(U)} \delta(U - H(\mathbf{q}, \mathbf{p})), \tag{33}$$

where $\delta(\cdot)$ is Dirac's delta function and $\omega(U)$ is the normalization

$$\omega(U) = \frac{1}{h^{3N}} \int_{\mathbb{R}^{3N}} \mathrm{d}\mathbf{p}_1 \cdots \mathrm{d}\mathbf{p}_N \int_{\mathbb{R}^{3N}} \mathrm{d}\mathbf{q}_1 \cdots \mathrm{d}\mathbf{q}_N \, \delta(U - H(\mathbf{q}, \mathbf{p})). \tag{34}$$

The constant h is introduced for dimensional reasons. Because of the quadratic nature of the kinetic energy term it is possible to integrate out the momenta $\mathbf{p}_j$. The resulting function is called the *configurational density function* and is evaluated as follows

$$\begin{aligned}
f_U^{\mathrm{conf}}(\mathbf{q}) &= \frac{1}{h^{3N}} \int_{\mathbb{R}^{3N}} \mathrm{d}\mathbf{p}_1 \cdots \mathrm{d}\mathbf{p}_N \, f_U(\mathbf{q}, \mathbf{p}) \\
&= \frac{1}{h^{3N}} \frac{1}{\omega(U)} \int_{\mathbb{R}^{3N}} \mathrm{d}\mathbf{p}_1 \cdots \mathrm{d}\mathbf{p}_N \, \delta(U - H(\mathbf{q}, \mathbf{p})) \\
&= \frac{1}{h^{3N}} \frac{1}{\omega(U)} \frac{\mathrm{d}}{\mathrm{d}U} \int_{\mathbb{R}^{3N}} \mathrm{d}\mathbf{p}_1 \cdots \mathrm{d}\mathbf{p}_N \, \Theta\left(U - \mathcal{V}(\mathbf{q}) - \frac{1}{2m} \sum_{j=1}^{N} |\mathbf{p}_j|^2\right) \\
&= \frac{1}{h^{3N}} \frac{1}{\omega(U)} (2m)^{3N/2} B(3N) \frac{\mathrm{d}}{\mathrm{d}U} [U - \mathcal{V}(\mathbf{q})]_+^{3N/2} \\
&= \frac{1}{2h^{3N}} \frac{3N}{\omega(U)} (2m)^{3N/2} B(3N) \, [U - \mathcal{V}(\mathbf{q})]_+^{\frac{3}{2}N-1}
\end{aligned}$$

$$= c_N \exp_q\left(-\alpha(\theta) - \theta\mathcal{V}(\mathbf{q})\right), \tag{35}$$

with

$$\begin{aligned} c_N &= \left(\frac{2m}{h^2}\right)^{3N/2}, \qquad \theta = \frac{1}{1-q}\frac{1}{[\Gamma(3N/2)\omega(U)]^{1-q}}, \\ \alpha(\theta) &= \frac{3}{2}N - 1 - \theta U, \qquad q = 1 - \frac{2}{3N-2}. \end{aligned} \tag{36}$$

Assume now that $\omega(U)$ is a strictly increasing function. Then it can be inverted to obtain U as a function of θ. Under this assumption the configurational density function belongs to the q-exponential family. As a consequence, one immediately knows that $f_U^{\rm conf}(\mathbf{q})$ maximizes the entropy functional $I_q(f)$. It is then tempting to identify $I_q(f_U^{\rm conf})$ with the configurational entropy. However, as noted before, $\xi(I_q(f_U^{\rm conf}))$ could serve as well, where $\xi(u)$ is an arbitrary monotonic function.

9. On the definition of temperature

There is no consensus in the literature what is the correct definition of the thermodynamic entropy $S(U)$ for isolated systems. The question is of importance because it directly determines the definition of the thermodynamic temperature via the relation (10). Most often one quotes the definition known as Boltzmann's entropy

$$S(U) = k_B \ln \omega(U), \tag{37}$$

where $\omega(U)$ is the density of states (34). However, this choice of definition of entropy has some drawbacks. For instance, for the pendulum the entropy $S(U)$ as a function of internal energy U is a piecewise convex function instead of a concave function [2]. The lack of concavity can be interpreted as a microcanonical instability [29, 30]. But there is no physical reason why the pendulum should be classified as being instable at all energies.

The shortcomings of Boltzmann's entropy have been noticed long ago. A slightly different definition of entropy is [28, 31, 32, 33] (see also in [34] the reference to the work of A. Schlüter)

$$S(U) = k_B \ln \Omega(U), \tag{38}$$

where $\Omega(U)$ is the integral of $\omega(U)$ and is given by

$$\Omega(U) = \frac{1}{h^{3N}} \int_{\mathbb{R}^{3N}} \mathrm{d}\mathbf{p}_1 \cdots \mathrm{d}\mathbf{p}_N \int_{\mathbb{R}^{3N}} \mathrm{d}\mathbf{q}_1 \cdots \mathrm{d}\mathbf{q}_N \Theta(U - H(\mathbf{q}, \mathbf{p})). \tag{39}$$

An immediate advantage of (38) is that the resulting expression for the temperature T, defined by the thermodynamical formula

$$\frac{1}{T} = \frac{\mathrm{d}S}{\mathrm{d}U}, \tag{40}$$

coincides with the notion of temperature as used by experimentalists. Indeed, one finds

$$k_B T = \frac{\Omega(U)}{\omega(U)}. \tag{41}$$

For a harmonic oscillator the density of states $\omega(U)$ is a constant. Hence, (41) implies $k_B T = U$, as wanted. It is well-known that for classical monoatomic gases the r.h.s. of (41) coincides with twice the average kinetic energy per degree of freedom. Its significance is that the equipartition theorem, assigning $(1/2)k_BT$ to each degree of freedom, does hold for the kinetic energy also in the microcanonical ensemble. Quite often the average kinetic energy per degree of freedom is experimentally accessible and provides a unique way to measure accurately the temperature of the system.

But also (38) and (41) are subject to criticism. In small systems finite size corrections appear [34, 35] for a number of reasons. As argued in [35], the problem is not the equipartition of the kinetic energy over the various degrees of freedom, but the relation between temperature and kinetic energy.

10. Example

Let us follow [26] and consider the example of $3N$ non-interacting harmonic oscillators. The potential energy equals

$$\mathcal{V}(\mathbf{q}) = \frac{1}{2}m\omega^2\sum_{j=1}^{3N}\mathbf{q}_j^2. \tag{42}$$

One calculates

$$\begin{aligned}\Omega(U) &= \frac{1}{h^{3N}}\int_{\mathbb{R}^{3N}}\mathrm{d}\mathbf{p}_1\cdots\mathrm{d}\mathbf{p}_{3N}\int_{\mathbb{R}^{3N}}\mathrm{d}\mathbf{q}_1\cdots\mathrm{d}\mathbf{q}_{3N}\Theta\left(U-\frac{1}{2m}\sum_j\mathbf{p}_j^2-\frac{1}{2}m\omega^2\sum_j\mathbf{q}_j^2\right)\\ &= \frac{1}{(3N)!\,\omega^{3N}}\left(\frac{4\pi U}{h}\right)^{3N}.\end{aligned} \tag{43}$$

From (36) and (43) now follows

$$\frac{1}{\theta} = (1-q)\left[\frac{\Gamma(3N/2)}{\Gamma(3N)}\frac{1}{\omega^{3N}}\left(\frac{4\pi}{h}\right)^{3N}U^{3N-1}\right]^{1-q}. \tag{44}$$

Using Stirling's approximation one obtains

$$\frac{\Gamma(3N/2)}{\Gamma(3N)} \sim \sqrt{2}\left(\frac{\mathrm{e}}{6N}\right)^{3N/2} \tag{45}$$

so that

$$\frac{1}{\theta} \sim \mathrm{e}\left(\frac{4\pi}{h\omega}\frac{U}{3N}\right)^2, \tag{46}$$

which shows that $1/\theta$ is roughly proportional to U^2. This is not what one expects. In the canonical ensemble is the temperature $1/\beta$ proportional to U

$$\frac{1}{\beta} = \frac{U}{3N}. \tag{47}$$

One therefore concludes that the parameter θ cannot be the inverse temperature β. Note that $\theta = \mathrm{d}\tilde{S}^{\mathrm{conf}}/\mathrm{d}U^{\mathrm{conf}}$ holds with $\tilde{S}^{\mathrm{conf}} \equiv I_q(f_U^{\mathrm{conf}})$ and $U^{\mathrm{conf}} = \mathbb{E}_\theta\mathcal{V} = \frac{1}{2}U$. Therefore one concludes that $\tilde{S}^{\mathrm{conf}}$ cannot be the thermodynamic configurational entropy S^{conf}, but the two entropies must be related by a non-trivial function $\xi(u)$. From (47) follows

$$S^{\mathrm{conf}} = \frac{3N}{2}\ln U^{\mathrm{conf}} + A, \tag{48}$$

for some constant A. But (46) implies that

$$\tilde{S}^{\rm conf} \sim -\frac{1}{\rm e}\left(\frac{h\omega}{4\pi}\right)^2 \frac{(3N)^2}{U^{\rm conf}} + \text{constant}. \tag{49}$$

Therefore the relation between $S^{\rm conf}$ and $\tilde{S}^{\rm conf}$ is logarithmic. More precisely, $S^{\rm conf} = \xi(\tilde{S}^{\rm conf})$, with $\xi(x)$ of the form $\xi(x) = -3N\ln(B-x) + C$, with constants B and C. This suggests that Rényi's entropy functional is the right one to start with. Indeed, let $\alpha = 2-q$. The relation between Rényi's $I_\alpha(f)$ and $I_q(f)$, as given by (19), is $I_\alpha(f) = \xi(I_q(f))$ with

$$\xi(u) = -\frac{1}{1-q}\ln\left(1-(1-q)u\right). \tag{50}$$

Take the constant in (49) equal to $1/(1-q)$. Then one obtains

$$S^{\rm conf} \equiv \xi\left(\tilde{S}^{\rm conf}\right) \simeq \left(\frac{3N}{2}-1\right)\left[\ln\frac{U^{\rm conf}}{6N} + \text{ constant}\right]. \tag{51}$$

This yields

$$\frac{{\rm d}S^{\rm conf}}{{\rm d}U^{\rm conf}} = \left(\frac{3N}{2}-1\right)\frac{1}{U^{\rm conf}} = \frac{3N-2}{U}, \tag{52}$$

which is an acceptable expression for the inverse temperature β.

11. Concluding remarks

The notion of a q-exponential family of statistical models has been reviewed. The q-Gaussian distribution belongs to this family. The probability distribution of the momentum of a single particle in a classical gas is such a q-Gaussian. An immediate consequence is that the velocity distribution of a classical particle is a q-deformed Maxwell distribution. In particular, the standard Maxwell distribution is an approximation which is only valid in the limit of large systems.

The configurational probability distribution of a classical gas always belongs to the q-exponential family. The non-extensivity parameter q is given by

$$\frac{1}{1-q} = \frac{3}{2}N - 1, \tag{53}$$

where N is the number of particles. The latter expression has appeared quite often in the literature, see for instance [27, 36, 37].

Finally, a system of $3N$ independent harmonic oscillators is considered. It is known that Tsallis' entropy and Rényi's entropy are equivalent in the context of the maximal entropy principle. But they make a difference from a thermodynamical point of view. The example suggests that the correct thermodynamical definition of temperature requires the use of Rényi's entropy. This observation has been made before in a different setting [28, 33].

References

[1] Naudts J 2004 *J. Ineq. Pure Appl. Math.* **5** 102 (*Preprint* `arXiv:math-ph/0402005`)
[2] Naudts J 2005 *Europhys. Lett.* **69** 719–724 (*Preprint* `arXiv:cond-mat/0412683`)
[3] Naudts J 2005 *Open Systems and Information Dynamics* **12** 13–22
[4] Naudts J 2006 *Physica A* **365** 42–49
[5] Naudts J 2008 *Entropy* **10** 131–149

[6] Naudts J 2009 *Cent. Eur. J. Phys.* **7** 405–413
[7] Grünwald P D and Dawid A P 2004 *Ann. Stat.* **32** 1367
[8] Eguchi S 2006 *Sugaku Expositions* **19** 197–216
[9] Sears T D and Vishwanathan S V N *Fenchel duality and Generalized maxent* preprint
[10] Abe S 2003 *J. Phys. A* **36** 8733–8738
[11] Hanel R and Thurner S 2007 *Physica A* **380** 109–114
[12] Topsøe F 2007 (*AIP Conference Proceedings* vol 965) pp 104–113
[13] Topsøe F 2009 *J. Global Optim.* **43** 553–564
[14] Amari S 1985 *Differential-geometrical methods in statistics* (*Lecture Notes in Statistics* vol 28) (New York, Berlin: Springer)
[15] Ohara A 2007 *Phys. Lett. A* **370** 184 – 193
[16] Ohara A 2009 *Eur. Phys. J. B* **70** 15–28
[17] Callen H 1985 *Thermodynamics and an Introduction to Thermostatistics, 2nd edn* (New York: Wiley)
[18] Tsallis C 2009 *Introduction to nonextensive statistical mechanics* (Springer)
[19] Tsallis C 1994 *Quimica Nova* **17** 468
[20] Naudts J 2002 *Physica A* **316** 323–334 (*Preprint* `cond-mat/0203489`)
[21] Tsallis C 1988 *J. Stat. Phys.* **52** 479–487
[22] Vignat C and Plastino A 2007 *J. Phys. A* **40** F969–F978
[23] Vignat C and Plastino A 2007 *Phys. Lett. A* **365** 370–375
[24] Ohara A and Wada T 2008 (*Preprint* `arXiv:0810.0624`)
[25] Vignat C and Plastino A 2005 *Phys. Lett. A* **343** 411–416
[26] Naudts J and Baeten M 2009 *Entropy* **11** 285–294
[27] Plastino A R and Plastino A 1994 *Phys. Lett. A* **193** 140–143
[28] Campisi M 2007 *Physica A* **385** 501–517
[29] Gross D 2001 *Microcanonical Thermodynamics: Phase transitions in 'small' systems* (*Lecture Notes in Physics* vol 66) (World Scientific)
[30] Gross D 1990 *Rep.Progr.Phys.* **53** 605–658
[31] Schlüter A 1948 *Z. Naturforschg.* **3a** 350–360
[32] Pearson E M, Halicioglu T and Tiller W A 1985 *Phys. Rev. A* **32** 3030–3039
[33] Campisi M and Bagci G B 2007 *Phys. Lett. A* **362** 11–15 (*Preprint* `arXiv:cond-mat/0605089v2`)
[34] Shirts R B, Burt S R and Johnson A M 2006 *J. Chem. Phys.* **125** 164102
[35] Uline M J, Siderius D W and Cortib D S 2008 *J. Chem Phys.* **128** 124301
[36] Abe S, Martinez S, Pennini F and Plastino A 2001 *Phys. Lett. A* **278** 249–254
[37] Adiba A B, Moreirab A A, Andrade Jr J S and Almeida M P 2003 *Physica A* **322** 276–284

Journal of Physics: Conference Series **201** (2010) 012004 doi:10.1088/1742-6596/201/1/012004

Nonlinear Fokker-Planck Equations Associated with Generalized Entropies: Dynamical Characterization and Stability Analyses

Masatoshi Shiino

Department of Applied Physics, Faculty of Science,
Tokyo Institute of Technology, Meguro-ku, Tokyo, Japan

E-mail: mshiino@ap.titech.ac.jp

Abstract. Dynamical characterization and behavior of two types of nonlinear Fokker-Planck equations (NFPEs) are studied within the framework of generalized thermostatistics. On the basis of generalized entropies NFPEs are shown to be constructed so that H-theorems hold with Lyapunov functionals that are given by free energy functionals associated with the generalized entropies. In the case of the ordinary type of NFPEs such H-theorems ensure convergence of solutions to their uniquely determined equilibrium solutions. In the case of mean-field type NFPEs (DNFPEs) that may exhibit bifurcation phenomena the H-theorems are shown to still hold to ensure global stability of solutions. Systematic description is given of local stability analysis based on the second-order variations of the free energy functionals.

1. Introduction

In recent years nonlinear Fokker-Planck equations (NFPEs) [1] have been attracting much attention in statistical physics as well as in its related fields of science such as computational neuroscience. While Fokker-Planck equations [2] are well known to arise from the so-called master equations of Markovian stochastic systems exhibiting diffusion processes and then take the form of linear equations in the probability density, NFPEs appear within different framework. The latter may be classified according to how the nonlinearity involved manifests itself, and also be differently classified into two types of categories according to whether bifurcations of solutions occur or not.

NFPEs without bifurcations are studied mostly in connection with the concept of non-additive entropies that are often called generalized entropies. One example of the simplest version of such NFPEs was first proposed by Plastino and Plastino [5] within the framework of Tsallis statistics [6, 7]. Their NFPE allows an H-theorem [4, 20, 21] to hold with the Lyapunov functional that is given by the free energy functional defined based on the use of the Tsallis entropy. The H-theorem there plays an important role and ensures convergence to a uniquely determined equilibrium solution (probability density), implying that the free energy continues to decrease until equilibrium state is reached [4, 20, 21]. Alternatively speaking, one can no longer expect the occurrence of bifurcations of solutions. The present author also studied previously the same kind of issue for the NFPE [21] associated with the Sharma -Mittal entropy [3] (see also [8, 22]). It is of interest to study systematically the relationship between generalized entropies and the associated NFPEs in terms of free energies with which H-theorems hold. The result of the study undertaken from the viewpoint of dynamical characterization of NFPEs was briefly reported in [26] . The similar problem was also studied with the use of an alternative systematic method involving the irreversible thermodynamic theory and the probability current [1].

Another type of NFPEs are mean-field type NFPEs where the nonlinearity arises from the concept of mean-field couplings together with taking the thermodynamic limit [9, 10, 11, 12, 13, 14, 15, 16, 17, 18, 19, 1], a combination of which may lead to the occurrence of bifurcation phenomena. When the concept of free energy makes sense, an H-theorem also holds with the Lyapunov functional that is given by the free energy functional. The H-theorem , however, ensures global stability alone and local stability analysis is required to examine stability switching among bifurcated solutions, by computing the second-order variation of the free energy. A typical example was shown by the present author, within the framework of Gibbs-Boltzmann statistics, for an NFPE with a double well potential that can describe ferro-para magnetic phase transitions of mean-field type [11, 12].

From the viewpoint of both extending such an H-theorem as used for systems exhibiting bifurcations due to the mean-field concept to that for the case of NFPEs associated with generalized entropies and trying to observe the occurrence of bifurcations within the context of Tsallis thermostatistics, I have recently proposed double nonlinear Fokker-Planck equations (DNFPEs) of several types [23, 24, 25] and conducted global as well as local stability analyses .

In this article I begin with reproducing the result of systematic study of the relationship between free energies based on generalized entropies and generalized NFPEs, that is described only briefly in a previous paper [26]. I present a general and systematic formulation of stability analyses of DNFPEs associated with a wide class of generalized entropies.

2. Generalized entropies together with free energies characterize nonlinear Fokker-Planck equations

We let the real-valued functions S and $w(\geq 0)$ satisfy the following conditions

$$(1)\ S^{''} \leq 0, \tag{1}$$

$$(2)\ S^{'} \geq 0,\ w^{''} \leq 0\ ,or\ \ S^{'} \leq 0,\ w^{''} \geq 0, \tag{2}$$

where the prime and double prime represent respectively the first and second order derivative of the functions. Assuming a generalized entropy to take the form

$$S = S\left[\int w(p)dx\right], \tag{3}$$

where $p(x)$ denotes a probability density and the bracket [ ] the argument of the variable, we define the free energy as usual

$$F \ = U \ - \gamma S \ = \ \int p(x)\ \varphi(x)\ dx \ - \gamma S\left[\int w(p(x))\ dx\right], \tag{4}$$

with γ being a positive constant and U representing the energy given by the average of potential φ:

$$U = \int p\varphi(x)dx. \tag{5}$$

We are concerned with a nonlinear Fokker Planck equation (NFPE) that reads

$$\frac{\partial\, p(t,x)}{\partial\, t} \ = - \ \frac{\partial}{\partial\, x}\Big(- \ \varphi^{'}(x)p(t,x)\Big) \ + \ \frac{\partial^2}{\partial\, x^2}K(p(t,x)), \tag{6}$$

Journal of Physics: Conference Series **201** (2010) 012004 doi:10.1088/1742-6596/201/1/012004

where the probability density $p(t, x)$ is defined for $-\infty < x < \infty$ and $K(p)$ denotes a certain positive-valued nonlinear function of p. The $K(p)$ is to be determined so that the free energy F monotonically decreases with time when the time evolution equation of $p(t, x)$ is given by the NFPE(6). Assuming the natural boundary condition so that the probability current

$$j = -\varphi'(x)p - \frac{\partial}{\partial x}K(p) \tag{7}$$

vanishes at $x = \pm\infty$, the equilibrium probability density given by $j = 0$ satisfies

$$\varphi' p_{\rm eq} = -\frac{\partial}{\partial x}K(p_{\rm eq}). \tag{8}$$

Assuming $p(t, x)$ to obey the above NFPE we compute the time derivative of the free energy F:

$$\begin{aligned}\frac{dF}{dt} &= \int \varphi\frac{\partial p}{\partial t}dx - \gamma S'\left[\int w(p)\,dx\right]\int w'(p)\frac{\partial p}{\partial t}dx \\ &= \int dx\varphi\frac{\partial}{\partial x}\left(\varphi' p + \frac{\partial}{\partial x}K(p)\right) - \gamma S'\int dx w'(p)\frac{\partial}{\partial x}\left(\varphi' p + \frac{\partial}{\partial x}K(p)\right) \\ &= -\int dx\,\varphi'\left(\varphi' p + K'(p)\frac{\partial p}{\partial x}\right) + \gamma S'\int dx w''(p)\frac{\partial p}{\partial x}\frac{\partial}{\partial x}\left(\varphi' p + K'(p)\frac{\partial p}{\partial x}\right).\end{aligned} \tag{9}$$

Noting

$$\varphi' = -\frac{K'(p_{\rm eq})}{p_{\rm eq}}\frac{\partial p_{\rm eq}}{\partial x}, \tag{10}$$

we further rewrite $\frac{dF}{dt}$ as

$$\frac{dF}{dt} = \int I\left(\frac{\partial p}{\partial x}, \frac{\partial p_{\rm eq}}{\partial x}\right)\,dx, \tag{11}$$

with

$$\begin{aligned}I\left(\frac{\partial p}{\partial x}, \frac{\partial p_{\rm eq}}{\partial x}\right) &= -\left(K'(p_{\rm eq})\right)^2\frac{p}{p_{\rm eq}^2}\left(\frac{\partial p_{\rm eq}}{\partial x}\right)^2 + \gamma\,S'w''(p)K'(p)\left(\frac{\partial p}{\partial x}\right)^2 \\ &\quad + \frac{K'(p_{\rm eq})}{p_{\rm eq}}\left(K'(p) - \gamma\,S'w''(p)p\right)\frac{\partial p_{\rm eq}}{\partial x}\frac{\partial p}{\partial x}.\end{aligned} \tag{12}$$

The discriminant of the above quadratic form $I\left(\frac{\partial p}{\partial x}, \frac{\partial p_{\rm eq}}{\partial x}\right)$ is given by

$$D_{\rm isc} = \left[\frac{K'(p_{\rm eq})}{p_{\rm eq}}\right]^2\left(K'(p) + \gamma S'w''(p)p\right)^2. \tag{13}$$

Then the condition for the quadratic form I to be nonpositive definite can be given by $D_{\rm isc} \le 0$ and thus one has

$$K'(p) + \gamma S'w''(p)p = 0. \tag{14}$$

Integrating this one obtains $K(p)$:

$$K(p) = -\gamma S'\int^p w''(p)pdp. \tag{15}$$

Substituting eq.(15) into eq.(12) one has

$$
\begin{aligned}
I &= -p\left\{\frac{K'(p_{\rm eq})}{p_{\rm eq}}\frac{\partial p_{\rm eq}}{\partial x}-\frac{K'(p)}{p}\frac{\partial p}{\partial x}\right\}^2 \\
&= -\gamma^2 p\left\{\frac{\partial}{\partial x}\left(S' w'(p)-S'_{\rm eq}w'(p_{\rm eq})\right)\right\}^2. \qquad (16)
\end{aligned}
$$

Then it follows

$$\frac{dF}{dt}=-\gamma^2\int p\left\{\frac{\partial}{\partial x}\left(S' w'(p)-S'_{\rm eq}w'(p_{\rm eq})\right)\right\}^2 dx\le 0, \qquad (17)$$

where equality holds when

$$S' w'(p)-S'_{\rm eq}w'(p_{\rm eq})=c\ (const). \qquad (18)$$

Inequality (17) shows that an H-theorem of the standard type holds with the Lyapunov functional taking the form of free energy F of eq.(4). Substituting eq. (14) into eq.(10) , we obtain

$$\varphi' = \gamma S'_{\rm eq}w''(p_{\rm eq})\frac{\partial p_{\rm eq}}{\partial x}=\gamma S'_{\rm eq}\frac{\partial}{\partial x}w'(p_{\rm eq}), \qquad (19)$$

and thus the equilibrium probability density $p_{\rm eq}(x)$ turns out to satisfy

$$\varphi=\gamma S'_{\rm eq}w'(p_{\rm eq})+c_0(\gamma), \qquad (20)$$

where the constant $c_0(\gamma)$ is determined by the normalization condition. We note that applying the MAX ENT principle leads to the above equation. Accordingly the free energy can be rewritten as

$$F=\gamma\left[\int p\ S'_{\rm eq}w'(p_{\rm eq})dx - S\left[\int w(p)\ dx\right]\right]+c_0(\gamma). \qquad (21)$$

Using this expression we can easily show that the free energy is bounded from below by its value $F_{\rm eq}$ attained for the equilibrium probability density $p_{\rm eq}$:

$$F_{\rm eq}=\gamma\left[\int p_{\rm eq}\ S'_{\rm eq}w'(p_{\rm eq})dx - S\left[\int w(p_{\rm eq})\ dx\right]\right]+c_0(\gamma). \qquad (22)$$

To see this we write

$$F(p)-F(p_{\rm eq})=\gamma\left[\int S'_{\rm eq}w'(p_{\rm eq})(p-p_{\rm eq})\,dx-\left\{S\left[\int w(p)dx\right]-S\left[\int w(p_{\rm eq})\ dx\right]\right\}\right]. \qquad (23)$$

Noting the condition assumed for the functions w and S (eqs.(1,2)), we have

$$\int (w(p)-w(p_{\rm eq}))\,dx \le \int w'(p_{\rm eq})(p-p_{\rm eq})\,dx,\ \left(w''\le 0\right), \qquad (24)$$

$$\int (w(p)-w(p_{\rm eq}))\,dx \ge \int w'(p_{\rm eq})\ (p-p_{\rm eq})\,dx,\ \left(w''\ge 0\right), \qquad (25)$$

$$S\left[\int w(p)dx\right]-S\left[\int w(p_{\rm eq})dx\right]\le S'\left[\int w(p_{\rm eq})\ dx\right]\int (w(p)-w(p_{\rm eq}))\,dx, \qquad (26)$$

and hence

$$
\begin{aligned}
S\left[\int w(p)dx\right]-S\left[\int w(p_{\rm eq})dx\right] &\le S'\left[\int w(p_{\rm eq})\ dx\right]\int (w(p)-w(p_{\rm eq}))\,dx \\
&\le S'_{\rm eq}\int w'(p_{\rm eq})\ (p-p_{\rm eq})\,dx. \qquad (27)
\end{aligned}
$$

Accordingly substituting eq.(27) into eq.(23) leads to

$$F\left(p\right)-F\left(p_{\text{eq}}\right)\geq 0. \tag{28}$$

Thermodynamic properties of the equilibrium free energy

We can show that the equilibrium free energy F_{eq} exhibits the Legendre transform structure and the thermodynamic relations hold as usual, when γ is viewed as temperature:

$$\frac{\partial F_{\text{eq}}}{\partial\gamma}=-S_{\text{eq}}, \tag{29}$$

$$\frac{\partial F_{\text{eq}}}{\partial h}=-\left\langle x\right\rangle\equiv-\int x\ p_{\text{eq}}dx. \tag{30}$$

In eqs.(29) and (30) the potential φ has been replaced by $\varphi - hx$, where h is viewed as the parameter conjugate to the variable x, which is, for example, supposed to play the role of magnetic field.

Examples of application of the present scheme based on the generalized entropies (3)

We now provide two simple cases that are related to the Tsallis entropy. First we set $S(z)=\frac{1-z}{q-1}$ together with $w(p)=p^q$ we then have the well-known Tsallis entropy

$$S=\frac{1-\int p^q dx}{q-1}, \tag{31}$$

and the associated NFPE (6) with $K(p)=\gamma p^q$ that was proposed by Plastino and Plastino [5] The H-theorem for this NFPE was shown in [20, 21, 4]. and is recovered in the present scheme. Second, we set $S(z)=\frac{q}{1-q}\left(1-z^{-\frac{1}{q}}\right)$ together with $w(p)=p^q$. We obtain the Sharma and Mittal entropy [3]

$$S=\frac{q}{1-q}\left(1-\left(\int p^q dx\right)^{-\frac{1}{q}}\right), \tag{32}$$

for which $K(p)=\gamma p^q(\int p^q dx)^{-\frac{q+1}{q}}$. The H-theorem for the NFPE with this $K(p)$ was studied in [21] in connection with the so-called escort probability density associated with Tsallis thermostatistics of the third choice [21, 8]. We see that the H-theorem is also recovered in the present scheme.

3. Mean-field type nonlinear Fokker-Planck equations and stability analyses

We consider another type of NFPE that involves, besides the nonlinearity of $K(p)$in eq.(6), nonlinearity associated with the so-called "mean-field type coupling" term. We may call such an NFPE double nonlinear Fokker -Planck equation (DNFPE) [23, 24, 25, 26] and are concerned with the following equation:

$$\frac{\partial p(t,x)}{\partial t}=-\frac{\partial}{\partial x}\left(\left(-\varphi'(x)+J\int xp(t,x)dx\right)p(t,x)\right)+\frac{\partial^2}{\partial x^2}K(p(t,x)), \tag{33}$$

with

$$K(p)=-DS'\int^p w''(p)pdp, \tag{34}$$

where J and D represent positive constants and $J\int xp(t,x)dx$ is the mean-field type coupling term. The integration is assumed to be over $-\infty < x < \infty$. The above-mentioned mean-field type coupling term is well known to be responsible for the occurrence of bifurcation phenomena [9, 10, 11, 12], when the standard Boltzmann entropy is taken with $K(p) = Dp$ and the potential $\varphi(x)$, for example, double well type. We define the free energy as usual

$$\begin{aligned} F(p(\cdot)) &= U - DS \\ &= \int \varphi p dx - \frac{J}{2}\Big(\int xpdx\Big)^2 - DS\Big[\int w(p)dx\Big] \\ &= \int \Big(\varphi - Jx\int xpdx\Big)pdx + \frac{J}{2}\Big(\int xpdx\Big)^2 - DS\Big[\int w(p)dx\Big], \end{aligned} \tag{35}$$

where the bracket [ ] denotes the argument of the function. With this free energy functional taken as a Liapunov functional (H-functional) we can show an H-theorem for the DNFPE(33) [23, 24, 25, 26].

H-theorem :
Let $p(t,x)$ satisfy the DNFPE(33). Then we have
(1) the free energy $F((p(t,\cdot))$ is bounded from below:

$$F((p(t,\cdot)) > c_{\rm F}(const), \tag{36}$$

(2) $F((p(t,\cdot))$ is decreasing with time :

$$\frac{dF\,(p(t,\cdot))}{dt} \le 0. \tag{37}$$

The boundedness inequality (36) can be proved in the same way as in the case of [23]. The second inequality (37) can be easily obtained by a straightforward calculation by introducing $R_p(x)$ that is normalized and satisfies

$$\Big(\varphi' - J\langle x\rangle_p\Big)R_p = -\frac{\partial}{\partial x}K\left(R_p\right), \tag{38}$$

where

$$\langle x\rangle_p = \int xpdx. \tag{39}$$

Here we note that the $R_p(x)$ is uniquely determined when p is given, as in [21]. Noting eq.(38) instead of eq.(10) and repeating the procedure of obtaining eq.(17) one has

$$\begin{aligned} \frac{dF}{dt} &= \int (\varphi - Jx\langle x\rangle)\frac{\partial p}{\partial t}dx - DS'\int w'(p)\frac{\partial p}{\partial t}dx \\ &= -D^2\int p\Big\{\frac{\partial}{\partial x}\Big(S'w'(p) - S'_{R_p}w'(R_p)\Big)\Big\}^2 dx \le 0, \end{aligned} \tag{40}$$

where $S'_{R_p} = S'\left[\int w(R_p)dx\right]$ with the bracket [ ] denoting the argument of the function, and equality holds for $S'w'(p) - S'_{R_p}w'(R_p) = c\ (const)$. We note that equilibrium solutions correspond to $R_{p_{\rm eq}} = p_{\rm eq}$, which takes the form of a self-consistent equation for the equilibrium solutions. It is also noted that one can no longer expect uniqueness of the equilibrium density $P_{\rm eq}(x)$, since the self- consistent equation, in general, admit multi solutions. Which of those multi- solutions are relevant has to be determined by the stability condition. In other words,

there may occur bifurcation phenomena involving stability switches , as the control parameter D is varied.
The H-theorem here ensures global stability of the system such that uniquely existing equilibrium solution is stable.

To conduct local stability analysis [12, 23, 24, 25, 26]. we expand F around p_{eq}

$$\delta F \equiv F\left(p_{\text{eq}} + \delta p\right) - F\left(p_{\text{eq}}\right) = \delta^{(1)} F\left[\delta p\right] + \delta^{(2)} F\left[\delta p, \delta p\right] + ... \tag{41}$$

Noting

$$\int \delta p dx = 0, \tag{42}$$

and differentiating F with respect to p yields

$$\delta^{(1)} F = \int \left(DS'_{R_p} w'(R_p) - DS' w'(p)\right) \delta p dx, \tag{43}$$

where we noted

$$\varphi - J\left\langle x\right\rangle x = DS'_{R_p} w'(R_p), \tag{44}$$

that is derived from eqs.(34) and (38). $\delta^{(1)}F$ vanishes for $p = p_{\text{eq}}$. Differentiating F twice with respect to p yields

$$2\delta^2 F = D \int \left(S'_{R_p} w''(R_p)\delta R_p + w'(R_p) S''_{R_p} \int w'(R_p)\delta R_p dx - S' w''(p)\delta p - w'(p) S'' \int w'(p)\delta p dx\right) \delta p dx. \tag{45}$$

Differentiating eq.(44) with respect to p one obtains

$$-Jx \int x\delta p dx = DS'_{R_p} w''(R_p)\delta R_p + Dw'(R_p) S''_{R_p} \int w'(R_p)\delta R_p dx. \tag{46}$$

Multiplying the above equation by δp and integrating the product yields

$$-J\left(\int x\delta p dx\right)^2 = D \int S'_{R_p} w''(R_p)\delta R_p \delta p dx + D\ S''_{R_p} \int w'(R_p)\delta p\ dx \int w'(R_p)\delta R_p dx. \tag{47}$$

Substituting this into eq.(45) one obtains

$$\begin{aligned} 2\delta^2 F &= -J\left(\int x\delta p dx\right)^2 - DS' \int w''(p)(\delta p)^2 dx - D\left(\int w'(p)\delta p dx\right)^2 S'' \\ &= -J\left(\int x\delta p dx\right)^2 + D\left|S'\right| \int \left|w''(p)\right| (\delta p)^2 dx + D\left(\int w'(p)\delta p \text{dx}\right)^2 \left|S''\right|. \end{aligned} \tag{48}$$

The second line of R.H.S.of the above equation holds owing to the assumption for the functions S', S'', w'' eqs.(1) and (2).

Since the free energy F continues decreasing with time owing to the H-theorem, positive definiteness of $\delta^2 F$, i.e. $\delta^2 F > 0$ implies local stability of the p_{eq}. Following the standard procedure we evaluate the above second-order variation at $p = p_{\text{eq}}$. Assuming $\int x^2 \left|w''(p_{\text{eq}})\right|^{-1} dx < \infty$, $\int w'(p_{\text{eq}})^2 \left|w''(p_{\text{eq}})\right|^{-1} dx < \infty$, and $\xi(x) \in L^2$ (class of square integrable functions) we put [11, 12, 23, 24, 1, 19]

$$\delta p = \left|w''(p_{\text{eq}})\right|^{-\frac{1}{2}} \xi(x). \tag{49}$$

Applying the decomposition procedure in [11, 12, 23, 24, 1, 19] we write ξ as

$$\xi(x) = \left|w''(p_{\rm eq})\right|^{-\frac{1}{2}}\left(\alpha_1 + \alpha_2 x + \alpha_3 w'(p_{\rm eq})\right) + \xi_\perp, \tag{50}$$

where $\xi_\perp$ is perpendicular to $\xi - \xi_\perp$. Defining

$$I_0 = \int \left|w''(p_{\rm eq})\right|^{-1} dx,\ I_1 = \int x\left|w''(p_{\rm eq})\right|^{-1} dx,\ I_2 = \int x^2\left|w''(p_{\rm eq})\right|^{-1} dx, \tag{51}$$

$$J_0 = \int w'(p_{\rm eq})\left|w''(p_{\rm eq})\right|^{-1} dx,\ J_1 = \int w'(p_{\rm eq})x\left|w''(p_{\rm eq})\right|^{-1} dx,\ J_2 = \int w'(p_{\rm eq})^2\left|w''(p_{\rm eq})\right|^{-1} dx, \tag{52}$$

one has from eq.(42)

$$\alpha_1 I_0 + \alpha_2 I_1 + \alpha_3 J_0 = 0, \tag{53}$$

and from eq.(48)

$$2\delta^2 F = A{\alpha_2}^2 + 2B\alpha_2\alpha_3 + C{\alpha_3}^2, \tag{54}$$

with

$$A = \frac{D\left|S'\right|}{I_0}\left(-{I_1}^2 + I_0 I_2\right) - \frac{J}{{I_0}^2}\left(-{I_1}^2 + I_0 I_2\right)^2 + \frac{D\left|S''\right|}{{I_0}^2}(I_0 J_1 - J_0 I_1)^2, \tag{55}$$

$$B = (I_0 J_1 - J_0 I_1)\left\{\frac{D\left|S'\right|}{I_0} - \frac{J}{{I_0}^2}\left(-{I_1}^2 + I_0 I_2\right) + \frac{D\left|S''\right|}{{I_0}^2}\left(I_0 J_2 - {J_0}^2\right)\right\}, \tag{56}$$

$$C = \frac{D\left|S'\right|}{I_0}\left(I_0 J_2 - {J_0}^2\right) - \frac{J}{{I_0}^2}(I_0 J_1 - J_0 I_1)^2 + \frac{D\left|S''\right|}{{I_0}^2}\left(I_0 J_2 - {J_0}^2\right)^2, \tag{57}$$

where eq.(53) was used.

The stability condition of the equilibrium density is given by that the second-order variation of the free energy $\delta^2 F$ be positive definite. Hence the condition for this is the one that

$$A > 0, \tag{58}$$

and the discriminant of eq.(54) be non-positive:

$$B^2 - AC \leq 0. \tag{59}$$

Inequalities (58) and (59) yield the stability condition

$$v < \frac{u\,(u + rZ)}{Yu + r\,(YZ - X^2)}, \tag{60}$$

where

$$u = \left|S'\right|,\ v = \frac{J}{DI_0},\ r = \frac{\left|S''\right|}{I_0}, \tag{61}$$

$$X = I_0 J_1 - I_1 J_0,\ Y = I_0 I_2 - {I_1}^2,\ Z = I_0 J_2 - {J_0}^2. \tag{62}$$

It is noted that stability switches among multi-solutions to the self-consistent equation for the order parameter take place when $v = \frac{u(u+rZ)}{Yu+r(YZ-X^2)}$.

We show a simple example where the DNFPE is derived by adding the mean-field type coupling term to the Plastino and Plastino NFPE with $K(p) = Dp^q, 0 < q < 1$ that is described in the end of Sec.2. In this case one has

$$X = q|q(q-1)|^{-2}\left(\int p_{\mathrm{eq}}^{2-q}dx \int xp_{\mathrm{eq}}dx - \int xp_{\mathrm{eq}}^{2-q}dx\right),$$
$$Y = |q(q-1)|^{-2}\left(\int p_{\mathrm{eq}}^{2-q}dx \int x^2 p_{\mathrm{eq}}^{2-q}dx - \left(\int xp_{\mathrm{eq}}^{2-q}dx\right)^2\right),$$
$$Z = |(q-1)|^{-2}\left(\int p_{\mathrm{eq}}^{2-q}dx \int p_{\mathrm{eq}}^{q}dx - 1\right), \tag{63}$$

$$u = \frac{1}{|q-1|}, \ v = \frac{J}{D|q(q-1)|^{-1}\int p_{\mathrm{eq}}^{2-q}dx}, \ r = 0. \tag{64}$$

The stability condition (60) with substitution of the above mentioned parameters reads

$$1 - \frac{J}{Dq}\left(\int x^2 p_{\mathrm{eq}}^{2-q}dx - \frac{\left(\int xp_{\mathrm{eq}}^{2-q}dx\right)^2}{\int p_{\mathrm{eq}}^{2-q}dx}\right) > 0. \tag{65}$$

We see that the above result recovers the one previously obtained in [23].

4. Conclusions

We have explored dynamical behaviors of NFPEs with and without bifurcations within the framework of generalized thermostatistics based on generalized entropies of any kind including Tsallis one and Sharma and Mittal one. H-theorems together with the concept of free energies playing the role of Lyapunov functionals are a key ingredient to construct NFPEs : starting with providing a generalized entropy to define the associated free energy in a standard manner we construct the corresponding NFPE that exhibits an H-theorem with its Lyapunov functional given by the free energy. The H-theorem ensures convergence of any probability density given as a solution of the NFPE to a uniquely determined equilibrium density. In this case the NFPE has nothing to do with bifurcation phenomena . On the other hand, incorporating a simple version of the so-called mean-field type term into the above-mentioned NFPE to construct a DNFPE , we can in general expect the occurrence of bifurcations of its solutions. In such a case we have shown an H-theorem to still hold to ensure global stability of solutions of the DNFPE. For the purpose of studying stability exchanges associated with the occurrence of bifurcations local stability analysis becomes to be required. We have evaluated the second-order variation of the free energy functional and obtained the stability condition explicitly.

References

[1] Frank T D 2005 *Nonlinear Fokker-Planck Equations* (Springer-Verlag, Berlin)
[2] Risken H 1984 *The Fokker-Planck Equation* (Berlin: Springer-Verlag)
[3] Sharma B D and Mittal D P 1975 *J. Math. Sci.* **10** 28
[4] Frank T D and Daffertshofer A 2001 *Physica A* **295** 455
[5] Plastino A R and Plastino A 1995 *Physica A* **222** 347
[6] Tsallis C 1988 *J. Stat. Phys.* **52** 479
[7] Tsallis C, Mendes R S and Plastino A R 1998 *Physica A* **261** 534
[8] Frank T D and Plastino A R 2002 *Eur.Phys. J. B* **30** 543
[9] Desai R C and Zwanzig R 1978 *J. Stat. Phys.* **19** 1
[10] Dawson D A 1983 *J. Stat. Phys.* **31** 29
[11] Shiino M 1985 *Phys. Lett. A* **112** 302
[12] Shiino M 1987 *Phys. Rev. A* **36** 2393 ;

[13] Shiino M 1985 *Phys. Lett. A* **111** 396
[14] Sakaguchi H 1988 *Prog. Theor. Phys.* **79** 39
[15] Shiino M and Yoshida K 2001 *Phys. Rev. E* **63** 026210-1
[16] Ichiki A, Ito H and Shiino M 2007 *Physica E* **40** 402
[17] Ichiki A and Shiino M 2009 *EPL* **87** 30004
[18] Frank T D, Daffertshofer A, Peper C E, Beck P J, and Haken H 2001 *Physica D* **150** 219
[19] Frank T D 2005 *Phys. Rev. E* **72** 041703
[20] Shiino M 1999 *The 5th International Workshop on Similarity in Diversity* (1999) 63
[21] Shiino M 2001 *J. Math. Phys.* **42** 2540
[22] Frank T D and Daffertshofer A 2000 *Physica A* **285** 351
[23] Shiino M 2002 *J. Math. Phys.* **43** 2654
[24] Shiino M 2002 *J. K. Phys. Soc.* **40** 1037
[25] Shiino M 2003 *Phys. Rev. E* **67** 056118
[26] Shiino M 2003 Eds. Fujita S, Hara H, Morabito D, and Okamura Y 2003 *Similarity in Diversity* (Nova Science Publisher)

Finite difference and averaging operators in generalized entropies

T Wada[1], **A M Scarfone**[2]

[1] Department of Electrical and Electronic Engineering, Ibaraki University, Japan
[2] Consiglio Nazionale delle Ricerche (CNR–INFM) and Dipartimento di Fisica, Politecnico di Torino, Italy

E-mail: wada@mx.ibaraki.ac.jp

E-mail: antonio.scarfone@polito.it

Abstract. In the two-parameter generalization of thermostatistics based on the Sharma-Taneja-Mittal entropy not only the generalized entropic functional $S_{a,b}$ but also a new functional $\mathcal{I}_{a,b}$ plays a fundamental role. These functionals are related to the finite difference and averaging operators arising in finite difference calculus.

1. Introduction

In the last decades great attention has been devoted to a variety of generalized entropies [1] from which probability distributions with power-law tails can be derived by applying the Jaynes maximum entropy principle. Among the many, some researchers have studied the two parameter family of entropies $S_{a,b}$ given by

$$S_{a,b}[p] \equiv -\sum_i \frac{p_i^a - p_i^b}{a-b} = -\sum_i s_{a,b}(p_i) \ , \tag{1}$$

with real parameters a and b. $S_{a,b}$ is introduced for the first time by Sharma-Taneja [2] and Mittal [3] (STM-entropy), independently in the fields of information theory, and successively reconsidered by Borges-Roditi [4] and Kaniadakis-Lissia-Scarfone [5] in the fields of statistical physics. As firstly noted in [6] and later in [4], some generalized entropies can be obtained by acting finite-difference operators on a generating function $g^p(s) \equiv \sum_i p_i^s$. This holds true also for the entropy $S_{a,b}$ that can be obtained by applying the two-parameter Chakrabarti and Jagannathan difference quotient operator $\mathrm{D}_s^{a,b}$ [7] (a two parameter generalization (36) of Jackson derivative) on the generating function $g^p(s)$.

As it has been shown in [8], not only the finite-difference quotient operator but also its associated averaging operator plays a fundamental role for characterizing generalized entropies. Acting this averaging operator on the generating function $g^p(s)$ produces the quantity

$$\mathcal{I}_{a,b}[p] \equiv \sum_i \frac{p_i^a + p_i^b}{2} = \sum_i \iota_{a,b}(p_i). \tag{2}$$

In Eqs. (1) and (2), we introduced, for the sake of notational simplicity, the two functions

$$s_{a,b}(x) \equiv \frac{x^a - x^b}{a-b} \underset{(a,\,b)\to(1,\,1)}{\longrightarrow} x\,\ln(x)\ , \tag{3}$$

$$\iota_{a,b}(x) \equiv \frac{x^a + x^b}{2} \underset{(a,\,b)\to(1,\,1)}{\longrightarrow} x\ . \tag{4}$$

Of course the above limitting values don't depend on the order of taking the limits. The functional $\mathcal{I}_{a,b}$ seems to play an important role in the formulation of statistical mechanics based on the entropy $S_{a,b}$. For instance, by considering two statistically independent composite systems $\mathrm{A}\cup\mathrm{B}$, with $p^{\mathrm{A}\cup\mathrm{B}} = p^{\mathrm{A}}\,p^{\mathrm{B}}$, the entropy $S_{a,b}$ can be factorized according to [9]

$$S_{a,b}\left[p^{\mathrm{A}\cup\mathrm{B}}\right] = S_{a,b}\left[p^{\mathrm{A}}\right]\,\mathcal{I}_{a,b}\left[p^{\mathrm{B}}\right] + \mathcal{I}_{a,b}\left[p^{\mathrm{A}}\right]\,S_{a,b}\left[p^{\mathrm{B}}\right]\ , \tag{5}$$

and similarly for $\mathcal{I}_{a,b}$

$$\mathcal{I}_{a,b}\left[p^{\mathrm{A}\cup\mathrm{B}}\right] = \mathcal{I}_{a,b}\left[p^{\mathrm{A}}\right]\,\mathcal{I}_{a,b}\left[p^{\mathrm{B}}\right] + \left(\frac{a-b}{2}\right)S_{a,b}\left[p^{\mathrm{A}}\right]\,S_{a,b}\left[p^{\mathrm{B}}\right]\ . \tag{6}$$

Moreover, the functional $\mathcal{I}_{a,b}$ appears repeatedly in the generalized thermodynamic relations among several quantities like, for example, the entropy $S_{a,b}$, the free-energy $F_{a,b}$ and the partition function $Z_{a,b}$ [10]. In the limit of $(a,\,b)\to(1,\,1)$, $\mathcal{I}_{a,b}$ reduces to unity, since from Eq. (4) we see that

$$\lim_{(a,\,b)\to(1,\,1)} \mathcal{I}_{a,b}[p] = \sum_i p_i = 1\ . \tag{7}$$

Consequently it disappears in the standard thermostatistics.

The paper is organized as follows. In the next section 2, we introduce briefly the generalized entropy $S_{a,b}$ and the associated Legendre structures by emphasizing the importance of the functional $\mathcal{I}_{a,b}$. The two-parameter finite difference and averaging operators are introduced in section 3, and then in section 4, we relate the functionals $S_{a,b}$ and $\mathcal{I}_{a,b}$ with the finite difference calculus on a one-dimensional lattice. The final section 5 is devoted to our conclusions.

2. Sharma-Taneja-Mittal entropy

As it has been shown in [5], the entropy $S_{a,b}$ can be rewritten in the form

$$S_{a,b}[p] = -\sum_i p_i \ln_{\{a,b\}}(p_i)\ , \tag{8}$$

where the two-parameter logarithmic function

$$\ln_{\{a,b\}}(x) \equiv \frac{x^{a-1} - x^{b-1}}{a-b} \underset{(a,\,b)\to(1,\,1)}{\longrightarrow} \ln(x)\ , \tag{9}$$

with real parameters a and b, fulfills the relations $\ln_{\{a,b\}}(x) = \ln_{\{b,a\}}(x) = -\ln_{\{2-a,2-b\}}(1/x)$. Formula (8) mimics the standard expression of the Boltzmann-Gibbs-Shannon entropy (recovered in the limit of $(a,\,b)\to(1,\,1)$) by replacing the ordinary logarithm with its two-parameter generalization.

Entropy $S_{a,b}$ includes some well known one-parameter generalized entropies as special cases. For example, setting $a=q$ and $b=1$, we recover the Tsallis entropy [11]

$$S_q[p]=\sum_i\frac{p_i^q-p_i}{1-q}\ , \tag{10}$$

by setting $a=q_{_\mathrm{A}}$ and $b=q_{_\mathrm{A}}^{-1}$, $S_{a,b}$ reduces to the Abe entropy [6]

$$S_{q_{_\mathrm{A}}}[p]=\sum_i\frac{p_i^{q_{_\mathrm{A}}^{-1}}-p_i^{q_{_\mathrm{A}}}}{q_{_\mathrm{A}}^{-1}-q_{_\mathrm{A}}}\ , \tag{11}$$

whilst, by setting $a=1-\kappa$ and $b=1+\kappa$, we obtain the Kaniadakis entropy [12]

$$S_\kappa[p]=\sum_i\frac{p_i^{1-\kappa}-p_i^{1+\kappa}}{2\,\kappa}\ . \tag{12}$$

Maximizing $S_{a,b}$, constrained by the normalization of probabilities and the linear mean energy, according to

$$\frac{\delta}{\delta p}\Big(S_{a,b}[p]-\gamma\sum_i p_i-\beta\,U[p]\Big)=0\ , \tag{13}$$

where

$$U[p]=\sum_i p_i\,E_i\ , \tag{14}$$

leads to the following probability distribution function

$$p_i^{\mathrm{ME}}=\alpha\,\exp_{_{\{a,b\}}}\left[-\frac{1}{\lambda}(\gamma+\beta\,E_i)\right]\ . \tag{15}$$

Here we introduced the generalized exponential $\exp_{_{\{a,b\}}}(x)$ as the inverse function of the generalized logarithm

$$\exp_{_{\{a,b\}}}\Big(\ln_{_{\{a,b\}}}(x)\Big)=\ln_{_{\{a,b\}}}\Big(\exp_{_{\{a,b\}}}(x)\Big)=x\ . \tag{16}$$

We note that, in general, the two-parameter exponential $\exp_{_{\{a,b\}}}(x)$ cannot be expressed in terms of elementally analytic functions. In Eq. (15), γ and β are the Lagrange multipliers associated to the two constraints, whilst α and λ depend on the parameters a and b throughout the relations

$$\alpha=\left(\frac{a}{b}\right)^{\frac{1}{b-a}}\ ,\quad \lambda=\left(\frac{a^{b-1}}{b^{a-1}}\right)^{\frac{1}{b-a}}\ . \tag{17}$$

They are related, each to the other, by means of

$$a\,\alpha^{a-1}=b\,\alpha^{b-1}=\lambda\ ,\quad \ln_{_{\{a,b\}}}\left(\frac{1}{\alpha}\right)=\frac{1}{\lambda}\ . \tag{18}$$

The Legendre structure concerning the two parameter generalized entropy $S_{a,b}$ has been advanced in [10]. In particular, the generalized free energy $F_{a,\,b}$ and partition function $Z_{a,\,b}$, have been introduced according to

$$F_{a,\,b} = -\frac{\mathcal{I}_{a,b} + \gamma - \left(\frac{a+b}{2} - 1\right)\beta U}{\left(\frac{a+b}{2}\right)\beta} \,, \tag{19}$$

$$\ln_{\{a,b\}}(Z_{a,\,b}) = \frac{a+b}{2}\left[\mathcal{I}_{a,b} + \gamma - \left(\frac{a+b}{2} - 1\right)\beta U\right] \,, \tag{20}$$

which satisfy formally the ordinary thermodynamical relations

$$F_{a,b} = U - \frac{1}{\beta} S_{a,b} \,, \qquad \frac{dF_{a,b}}{d(1/\beta)} = -S_{a,b} \,, \tag{21}$$

$$S_{a,b} = \ln_{\{a,b\}}(Z_{a,\,b}) + \beta U \,, \qquad \frac{d}{d\beta}\ln_{\{a,b\}}(Z_{a,\,b}) = -U \,. \tag{22}$$

In the limit of $(a,\,b) \to (1,\,1)$, these mathematical structures reduce to the standard thermodynamic relations, in particular we obtain the well known relations

$$F = -\frac{1+\gamma}{\beta}, \quad \ln(Z) = 1 + \gamma \,, \tag{23}$$

for the standard free energy and the partition function of thermal systems.

3. Finite difference quotient and averaging operators

For the sake of simplicity, let us consider an one dimensional space. As well known, there are several difference operators associated with the differential operator d/dx. For instance, the forward, central, and backward difference operators for equally spaced points are given by

$$\begin{aligned} \Delta_{+x} f(x) &= f(x+\epsilon) - f(x) \,, \\ \Delta_{x} f(x) &= f(x+\epsilon/2) - f(x-\epsilon/2) \,, \\ \Delta_{-x} f(x) &= f(x) - f(x-\epsilon) \,, \end{aligned} \tag{24}$$

respectively. *Difference quotient operators* are defined in terms of these quantities according to

$$\begin{aligned} \mathrm{D}_{+x} f(x) &\equiv \frac{\Delta_{+x}}{\epsilon} = \frac{f(x+\epsilon) - f(x)}{\epsilon} \,, && \text{forward} \\ \mathrm{D}_{x} f(x) &\equiv \frac{\Delta_{x}}{\epsilon} = \frac{f(x+\epsilon/2) - f(x-\epsilon/2)}{\epsilon} \,, && \text{central} \\ \mathrm{D}_{-x} f(x) &\equiv \frac{\Delta_{-x}}{\epsilon} = \frac{f(x) - f(x-\epsilon)}{\epsilon} \,. && \text{backward} \end{aligned} \tag{25}$$

All of these difference quotient operators reduce to the single differential operator d/dx in the limit of $\epsilon \to 0$. In short, the finite difference calculus has a richness of varieties. In addition to the difference quotient operators, one can consider also the related forward, central, and backword averaging operators given by

$$\begin{aligned} \mathrm{M}_{+x} f(x) &= \frac{f(x+\epsilon) + f(x)}{2} \,, \\ \mathrm{M}_{x} f(x) &= \frac{f(x+\epsilon/2) + f(x-\epsilon/2)}{2} \,, \\ \mathrm{M}_{-x} f(x) &= \frac{f(x) - f(x-\epsilon)}{2} \,, \end{aligned} \tag{26}$$

respectively. We note that all these quantities reduce to the identity operator $\hat{I}f(x) = f(x)$, in the $\epsilon \to 0$ limit. Therefore, in the finite difference calculus both averaging and difference quotient operators are important tools, whereas in the continuous calculus, only the differential operator d/dx is relevant.

Furthermore, for non-equally spaced points, averaging operators play an important role because it is *non-commutative* with respect to the corresponding difference quotient operator. In order to see this non-commutativity, let us consider the following difference quotient operator

$$\mathrm{D}_n f(x_n) = \frac{f(x_{n+1}) - f(x_n)}{\epsilon_n} , \tag{27}$$

and the corresponding averaging operator

$$\mathrm{M}_n f(x_n) = \frac{f(x_{n+1}) + f(x_n)}{2} , \tag{28}$$

for non-equally spaced points x_n, with the separation $\epsilon_n \equiv x_{n+1} - x_n$.

Applying first the averaging operator and then the difference quotient operator to a function $f(x_n)$ it leads to

$$\mathrm{D}_n\mathrm{M}_n f(x_n) = \mathrm{D}_n \frac{f(x_{n+1}) + f(x_n)}{2} = \frac{f(x_{n+2}) - f(x_{n+1})}{2\,\epsilon_n} + \frac{f(x_{n+1}) - f(x_n)}{2\,\epsilon_n} . \tag{29}$$

Next reversing the order of the operations, we obtain

$$\mathrm{M}_n\mathrm{D}_n f(x_n) = \mathrm{M}_n \frac{f(x_{n+1}) - f(x_n)}{\epsilon_n} = \frac{f(x_{n+2}) - f(x_{n+1})}{2\,\epsilon_{n+1}} + \frac{f(x_{n+1}) - f(x_n)}{2\,\epsilon_n} . \tag{30}$$

Comparing these results clearly show the non-commutativity of the two operators in a non-equally spaced lattice: $\mathrm{D}_n\mathrm{M}_n \neq \mathrm{M}_n\,\mathrm{D}_n$. The origin of this non-commutativity is in the different denominators (ϵ_n and ϵ_{n+1} respectively) of the first terms in the right hand sides of Eqs. (29) and (30). In the case of equally spaced points, all ϵ_n reduce to the same value ϵ, and consequently the averaging and difference quotient operators become commutative.

Finally, by considering the action of the operators D_n and M_n on the product of two functions, we can derive the following *Leibniz rules*

$$\mathrm{D}_n\left[f(x_n)\,h(x_n)\right] = \mathrm{D}_n f(x_n)\,\mathrm{M}_n h(x_n) + \mathrm{M}_n f(x_n)\,\mathrm{D}_n h(x_n) , \tag{31}$$

$$\mathrm{M}_n\left[f(x_n)\,h(x_n)\right] = \mathrm{M}_n f(x_n)\,\mathrm{M}_n h(x_n) + \left(\frac{\epsilon_n}{2}\right)^2 \mathrm{D}_n f(x_n)\,\mathrm{D}_n h(x_n) , \tag{32}$$

for arbitrary functions $f(x)$ and $h(x)$. These relations clearly show how these two operators are intimately related each into the other.

4. Relationship between generalized entropies and finite difference quotient operators

A possible way to obtain generalized entropies from difference quotient operators has been illustrated by Abe in [6] by introducing a generating function

$$g^p(s) \equiv \sum_i p_i^s . \tag{33}$$

For example, by applying the Jackson derivative

$$\mathrm{D}_x^q\, f(x) \equiv \frac{f(q\,x) - f(x)}{(q-1)\,x} , \tag{34}$$

to the generating function (33), the Tsallis entropy is obtained as follows

$$S_q[p] = -\mathrm{D}_s^q\, g^p(s)\Big|_{s=1} = \frac{1-\sum_i p_i^q}{q-1} \,. \tag{35}$$

In a similar way, by applying the two-parameter difference quotient operator introduced by Chakrabarti and Jagannathan in [7]

$$\mathrm{D}_x^{a,b} f(x) \equiv \frac{f(a\,x)-f(b\,x)}{(a-b)\,x} \,, \tag{36}$$

Borges and Roditi [4] have derived the STM-entropy according to

$$S_{a,b}[p] = -\mathrm{D}_s^{a,b}\, g^p(s)\Big|_{s=1} = -\sum_i \frac{p_i^a - p_i^b}{a-b} = -\sum_i s_{a,b}(p_i) \,. \tag{37}$$

All these entropic forms reduce, in a suitable limit, to the standard Boltzmann-Gibbs-Shannon entropy, which is obtained as

$$S[p] = -\frac{d}{ds} g^p(s)\Big|_{s=1} = -\sum_i p_i \ln(p_i) \,. \tag{38}$$

since, both Jackson derivative (for $q \to 1$) and Chakrabarti-Jagannathan derivative (for $(a,\,b) \to (1,\,1)$) reduce to the standard differential operator d/ds.

We remark that the operators D_s^q and $\mathrm{D}_s^{a,b}$ are generalizations of the difference quotient operator (27) in a non-equally spaced lattice. In fact, for instance, by setting $x_n = b\,x$, $x_{n+1} = a\,x$ in Eq. (27) (so that $\epsilon_n = (a-b)\,x$) we obtain

$$D_n f(x) = \frac{f(a\,x)-f(b\,x)}{(a-b)\,x} = \mathrm{D}_x^{a,b} f(x) \,. \tag{39}$$

In a similar way, we can emphasized the role of the functional $\mathcal{I}_{a,b}$ which can be obtained by applying the two-parameter averaging operator $\mathrm{M}_s^{a,b}$ to the same generating function $g^p(s)$

$$\mathcal{I}_{a,b}[p] = \mathrm{M}_s^{a,b} g^p(s)\Big|_{s=1} = \sum_i \frac{p_i^a + p_i^b}{2} = \sum_i \iota_{a,b}(p_i) \,. \tag{40}$$

Here, the averaging operator $\mathrm{M}_x^{a,b}$ is also the generalization of the averaging operator (28) in a non-equally spaced lattice.

4.1. Leibniz rule and the generalized additivity

As shown in [8], the generalized additivity of $S_{a,b}$ is rooted in the Leibniz product rule concerning with the corresponding difference quotient operator. For the Chakrabarti-Jagannathan derivative $\mathrm{D}_x^{a,b}$, the Leibniz rule can be written, according to Eq. (31), in the following symmetric form

$$\mathrm{D}_x^{a,b}\Big[f(x)\,h(x)\Big] = \mathrm{D}_x^{a,b} f(x)\ \ \mathrm{M}_x^{a,b} h(x) + \mathrm{M}_x^{a,b} f(x)\ \ \mathrm{D}_x^{a,b} h(x) \,. \tag{41}$$

Therefore, by acting $\mathrm{D}_s^{a,b}$ on the generating function $g^{p^{A\cup B}}(s) = \sum_{i,j} p_{ij}^s = \sum_i p_i^s \sum_j p_{j|i}^s$, we obtain

$$-\sum_{i,j} \mathrm{D}_s^{a,b} p_{ij}^s\Big|_{s=1} = -\sum_i \mathrm{D}_s^{a,b} p_i^s\Big|_{s=1} \sum_j \mathrm{M}_s^{a,b} p_{j|i}^s\Big|_{s=1} - \sum_i \mathrm{M}_s^{a,b} p_i^s\Big|_{s=1} \sum_j \mathrm{D}_s^{a,b} p_{j|i}^s\Big|_{s=1} \,. \tag{42}$$

By recalling the definitions (3) and (4), we can write

$$S_{a,b}[p^{\mathrm{A}\cup\mathrm{B}}] = \sum_i s_{a,b}(p_i) \sum_j \iota_{a,b}(p_{j|i}) + \sum_i \iota_{a,b}(p_i) \sum_j s_{a,b}(p_{j|i}) \ , \tag{43}$$

that is the two-parameter generalization of Shannon additivity [13, 14] for the STM-entropy. In the limit of $(a,\, b) \to (1,\, 1)$, this equation reduces to the standard Shannon additivity [13],

$$S[p^{\mathrm{A}\cup\mathrm{B}}] = S[p^{\mathrm{A}}] + \sum_i p_i \sum_j s(p_{j|i}) \ , \tag{44}$$

whilst, for statistically independent systems with $p_{ij} = p_i\, p_j$, it reduces to Eq. (5). Thus, the generalized additivity for the two-parameter generalized entropies $S_{a,b}$ is nothing but a consequence of the Leibniz rule for the two-parameter difference quotient operator $\mathrm{D}_s^{a,b}$.

In a similar way, from the relation

$$\mathrm{M}_x^{a,b}\Big[f(x)\, h(x)\Big] = \mathrm{M}_x^{a,b} f(x)\ \mathrm{M}_x^{a,b} h(x) + \left(\frac{a-b}{2}\right)^2 \mathrm{D}_x^{a,b} f(x)\ \mathrm{D}_x^{a,b} h(x) \ , \tag{45}$$

we can obtain the additivity rule for the functional $\mathcal{I}_{a,b}$ by the action of $\mathrm{M}_s^{a,b}$ on the generator $g^{p^{\mathrm{A}\cup\mathrm{B}}}(s)$,

$$\begin{aligned}\sum_{i,j} \mathrm{M}_s^{a,b}\, p_{ij}^s\Big|_{s=1} =& \sum_i \mathrm{M}_s^{a,b}\, p_i^s\Big|_{s=1} \sum_j \mathrm{M}_s^{a,b} p_{j|i}^s\Big|_{s=1} \\ &+ \left(\frac{a-b}{2}\right)^2 \sum_i \mathrm{D}_s^{a,b}\, p_i^s\Big|_{s=1} \sum_j \mathrm{D}_s^{a,b}\, p_{j|i}^s\Big|_{s=1} \ .\end{aligned} \tag{46}$$

As a consequence, we can derive the relation

$$\mathcal{I}_{a,b}[p^{\mathrm{A}\cup\mathrm{B}}] = \sum_i \iota_{a,b}(p_i) \sum_j \iota_{a,b}(p_{j|i}) + \left(\frac{a-b}{2}\right)^2 \sum_i s_{a,b}(p_i) \sum_j s_{a,b}(p_{j|i}) \ , \tag{47}$$

which reduces to Eq. (6) for statistically independent systems. In the limit of $(a,\, b) \to (1,\, 1)$, this last relation reduces to a merely trivial identity: $1 = 1$.

5. Conclusion

In the two-parameter generalization of thermostatistics based on the Sharma-Taneja-Mittal entropy, we have shown that not only the entropic functional $S_{a,b}$ but also the complementary functional $\mathcal{I}_{a,b}$ plays a fundamental role. As it has been shown, the functional $\mathcal{I}_{a,b}$ disappears in the limit of $(a,\, b) \to (1,\, 1)$. This fact suggest that a consistent formulation of the statistical mechanics for complex systems could require further "informations" carried out by the functional $\mathcal{I}_{a,b}$, with respect to the "ingredients" which are essential to describe ideal and non-interacting systems. Although the physical meaning of the functional $\mathcal{I}_{a,b}$ is still not yet clear, there are some evidences about its relevance in the developed of a generalized thermostatistics based on the entropic form $S_{a,b}$.

Acknowledgments

This research was partially supported by the Ministry of Education, Science, Sports and Culture (MEXT), Japan, Grant-in-Aid for Scientific Research (B), 18300003, and (C), 19540391, 2009.

References

[1] Kaniadakis G 2009 *Eur. Phys. J. B* **70** 3
[2] Sharma B D and Taneja L J 1975 *Metrika* **22** 205
[3] Mittal D P 1975 *Metrika* **22** 35
[4] Borges E P and Roditi I 1998 *Phys. Lett. A* **246** 399
[5] Kaniadakis G, Lissia M and Scarfone A M 2005 *Phys. Rev. E* **71** 046128
[6] Abe S 1997 *Phys. Lett. A* **224** 326
[7] Chakrabarti R and Jagannathan R, 1991 *J. Phys. A: Math. Gen.* **24** L711
[8] Wada T and Suyari H, 2007 *Phys. Lett. A* **368** 199-205
[9] Scarfone A M and Wada T, 2005 *Phys. Rev. E* **72** 026123
[10] Scarfone A M 2006 *Physica A* **365** 63
[11] Tsallis C 2009 *Introduction To Nonextensive Statistical Mechanics: Approaching A Complex World* (New York: Springer-Verlag)
[12] Kaniadakis G 2001 *Physica A* **296** 405; Kaniadakis G 2009 *Eur. Phys. J. A* **40** 275
[13] Shannon C E 1948 *Bell Syst. Tech. J* **24** 379
[14] Suyari H 2004 *IEEE Trans. Inform. Theory* **50** 1783

A statistical mechanical interpretation of algorithmic information theory: Total statistical mechanical interpretation based on physical argument

Kohtaro Tadaki

Research and Development Initiative, Chuo University, 1–13–27 Kasuga, Bunkyo-ku, Tokyo 112-8551, Japan

E-mail: tadaki@kc.chuo-u.ac.jp

Abstract. The statistical mechanical interpretation of algorithmic information theory (AIT, for short) was introduced and developed by our former works [K. Tadaki, Local Proceedings of CiE 2008, pp. 425–434, 2008] and [K. Tadaki, Proceedings of LFCS'09, Springer's LNCS, vol. 5407, pp. 422–440, 2009], where we introduced the notion of thermodynamic quantities, such as partition function $Z(T)$, free energy $F(T)$, energy $E(T)$, statistical mechanical entropy $S(T)$, and specific heat $C(T)$, into AIT. We then discovered that, in the interpretation, the temperature T equals to the partial randomness of the values of all these thermodynamic quantities, where the notion of partial randomness is a stronger representation of the compression rate by means of program-size complexity. Furthermore, we showed that this situation holds for the temperature T itself, which is one of the most typical thermodynamic quantities. Namely, we showed that, for each of the thermodynamic quantities $Z(T)$, $F(T)$, $E(T)$, and $S(T)$ above, the computability of its value at temperature T gives a sufficient condition for $T \in (0, 1)$ to satisfy the condition that the partial randomness of T equals to T. In this paper, based on a physical argument on the same level of mathematical strictness as normal statistical mechanics in physics, we develop a total statistical mechanical interpretation of AIT which actualizes a perfect correspondence to normal statistical mechanics. We do this by identifying a microcanonical ensemble in the framework of AIT. As a result, we clarify the statistical mechanical meaning of the thermodynamic quantities of AIT.

1. Introduction

Algorithmic information theory (AIT, for short) is a framework for applying information-theoretic and probabilistic ideas to recursive function theory. One of the primary concepts of AIT is the *program-size complexity* (or *Kolmogorov complexity*) $H(s)$ of a finite binary string s, which is defined as the length of the shortest binary input for a universal decoding algorithm U, called an *optimal computer*, to output s. By the definition, $H(s)$ is thought to represent the amount of randomness contained in a finite binary string s, which cannot be captured in a computational manner. In particular, the notion of program-size complexity plays a crucial role in characterizing the *randomness* of an infinite binary string, or equivalently, a real. In [4] Chaitin introduced the halting probability Ω as an example of random real. His Ω is defined as the probability that the optimal computer U halts, and plays a central role in the metamathematical development of AIT. The first n bits of the base-two expansion of Ω solve the halting problem of U for inputs of length at most n. By this property, the base-two expansion of Ω is shown to be a random infinite binary string.

In the works [10, 11] we introduced the notion of *partial randomness* for a real as a stronger representation of the compression rate of a real by means of program-size complexity. At the same time, we generalized Chaitin's halting probability Ω to $\Omega(D)$ by

$$\Omega(D) = \sum_{p \in \operatorname{dom} U} 2^{-\frac{|p|}{D}}, \tag{1}$$

so that the partial randomness of $\Omega(D)$ can be controlled by a real D with $0 < D \leq 1$.[1] Here, $\operatorname{dom} U$ denotes the set of all programs for U (i.e., the set of all halting inputs for U). As D becomes larger, the partial randomness of $\Omega(D)$ increases. When $D = 1$, $\Omega(D)$ becomes a random real, i.e., $\Omega(1) = \Omega$.

In 2006 Calude and Stay [3] pointed out a formal correspondence between $\Omega(D)$ and a partition function in statistical mechanics. In statistical mechanics, the partition function $Z_{\mathrm{sm}}(T)$ at temperature T is defined by

$$Z_{\mathrm{sm}}(T) = \sum_{x \in X} e^{-\frac{E_x}{k_{\mathrm{B}} T}}, \tag{2}$$

where X is a complete set of energy eigenstates of a statistical mechanical system and E_x is the energy of an energy eigenstate x. The positive constant k_{B} is called the *Boltzmann Constant*. Calude and Stay [3] pointed out, in essence, that the partition function $Z_{\mathrm{sm}}(T)$ has the same form as $\Omega(D)$ by performing the following replacements in $Z_{\mathrm{sm}}(T)$:

Replacements 1.1.

(i) Replace the complete set X of energy eigenstates x by the set $\operatorname{dom} U$ *of all programs p for U.*

(ii) Replace the energy E_x of an energy eigenstate x by the length $|p|$ of a program p.

(iii) Set the Boltzmann Constant k_{B} to $1/\ln 2$, where the ln *denotes the natural logarithm.* □

Inspired by this suggestion of Calude and Stay, starting from the works [13, 14] we have developed so far a statistical mechanical interpretation of AIT, where $\Omega(D)$ appears as a partition function.

Generally speaking, in order to give a statistical mechanical interpretation to a framework which looks unrelated to statistical mechanics at first glance, it is important to identify a *microcanonical ensemble* in the framework. Once we can do so, we can easily develop an equilibrium statistical mechanics on the framework according to the theoretical development of normal equilibrium statistical mechanics. Here, the microcanonical ensemble is a certain sort of uniform probability distribution. In fact, in the work [12] we developed a statistical mechanical interpretation of the noiseless source coding scheme in information theory by identifying a microcanonical ensemble in the scheme. Then, based on this identification, in [12] the notions in statistical mechanics such as statistical mechanical entropy, temperature, and thermal equilibrium are translated into the context of noiseless source coding.

Thus, in order to develop a total statistical mechanical interpretation of AIT, it is appropriate to identify a microcanonical ensemble in the framework of AIT. Note, however, that AIT is not a physical theory but a purely mathematical theory. Therefore, in order to obtain significant results for the development of AIT itself, we have to develop a statistical mechanical interpretation of AIT in a mathematically rigorous manner, unlike in normal statistical mechanics in physics where arguments are not necessarily mathematically rigorous. A fully rigorous mathematical treatment of statistical mechanics is already developed (see Ruelle [9]). At present, however, it would not as yet seem to be an easy task to merge AIT with this mathematical treatment in a satisfactory manner.

In our former works [13, 14], for mathematical strictness we developed a statistical mechanical interpretation of AIT in a different way from the idealism above. In [13] we introduced the notion of *thermodynamic quantities at temperature T*, such as partition function $Z(T)$, free energy $F(T)$, energy $E(T)$, statistical mechanical entropy $S(T)$, and specific heat $C(T)$, into AIT by performing Replacements 1.1 for the corresponding thermodynamic quantities at temperature T in statistical mechanics. These thermodynamic quantities in AIT are real functions of a real

[1] In [10, 11], $\Omega(D)$ is denoted by Ω^D.

argument $T > 0$. We prove that if the temperature T is a computable real with $0 < T < 1$ then, for each of these thermodynamic quantities, the partial randomness (and therefore the compression rate) of its value equals to T. Thus, the temperature T plays a role as the partial randomness of the values of the thermodynamic quantities in our statistical mechanical interpretation of AIT. Furthermore, we showed that this situation holds for the temperature T itself, which is one of the most typical thermodynamic quantities. Namely, we showed in [13, 14] that for each of the thermodynamic quantities $Z(T)$, $F(T)$, $E(T)$, and $S(T)$, the computability of its value at temperature T gives a sufficient condition for $T \in (0,1)$ to be a *fixed point on partial randomness* (see Theorems 3.4, 3.5, 3.6, and 3.7 below).

In this paper we show that, if we do not stick to the mathematical strictness of an argument, we can certainly develop a total statistical mechanical interpretation of AIT which attains a perfect correspondence to normal statistical mechanics. In the total interpretation, we identify a microcanonical ensemble in AIT in a similar manner to [12], based on the probability measure which gives Chaitin's Ω the meaning of the halting probability actually. This identification enables us to clarify the statistical mechanical meaning of the thermodynamic quantities of AIT, which are originally introduced by [13] in a rigorous manner.

The paper is organized as follows. We begin in Section 2 with some preliminaries to AIT and partial randomness. In Section 3, we review the main rigorous results on the statistical mechanical interpretation of AIT, obtained so far in our former works [13, 14]. In Section 4, based on a physical argument on the same level of mathematical strictness as normal statistical mechanics, we develop a total statistical mechanical interpretation of AIT. We conclude this paper with a mention to the future direction of this work in Section 5.

2. Preliminaries

We start with some notation about numbers and strings which will be used in this paper. $\mathbb{N} = \{0, 1, 2, 3, \dots\}$ is the set of natural numbers, and $\mathbb{N}^+$ is the set of positive integers. $\mathbb{Q}$ is the set of rationals, and $\mathbb{R}$ is the set of reals. $\{0,1\}^* = \{\lambda, 0, 1, 00, 01, 10, 11, 000, \dots\}$ is the set of finite binary strings, where λ denotes the *empty string*. For any $s \in \{0,1\}^*$, $|s|$ is the *length* of s. A subset S of $\{0,1\}^*$ is called *prefix-free* if no string in S is a prefix of another string in S. $\{0,1\}^\infty$ is the set of infinite binary strings, where an infinite binary string is infinite to the right but finite to the left. For any function f, the domain of definition of f is denoted by $\operatorname{dom} f$. Normally, $o(n)$ denotes any function $f\colon \mathbb{N}^+ \to \mathbb{R}$ such that $\lim_{n\to\infty} f(n)/n = 0$.

Let α be an arbitrary real. We denote by $\alpha\restriction_n \in \{0,1\}^*$ the first n bits of the base-two expansion of $\alpha - \lfloor \alpha \rfloor$ with infinitely many zeros, where $\lfloor \alpha \rfloor$ is the greatest integer less than or equal to α. For example, in the case of $\alpha = 5/8$, $\alpha\restriction_6 = 101000$. A real α is called *computable* if there exists a total recursive function $f\colon \mathbb{N}^+ \to \mathbb{Q}$ such that $|\alpha - f(n)| < 1/n$ for all $n \in \mathbb{N}^+$.

2.1. Algorithmic information theory

In the following we concisely review some definitions and results of AIT [4, 5]. A *computer* is a partial recursive function $C\colon \{0,1\}^* \to \{0,1\}^*$ such that $\operatorname{dom} C$ is a nonempty prefix-free set. For each computer C and each $s \in \{0,1\}^*$, $H_C(s)$ is defined by $H_C(s) = \min\left\{\, |p| \;\middle|\; p \in \{0,1\}^* \,\&\, C(p) = s \,\right\}$ (may be ∞). A computer U is said to be *optimal* if for each computer C there exists $d \in \mathbb{N}$ such that, for every $s \in \{0,1\}^*$, $H_U(s) \le H_C(s) + d$. It is easy to see that there exists an optimal computer. We choose a particular optimal computer U as the standard one for use, and define $H(s)$ as $H_U(s)$, which is referred to as the *program-size complexity* of s or the *Kolmogorov complexity* of s. A finite binary string in $\operatorname{dom} U$ is called a *program for* U.

For each $s \in \{0,1\}^*$, $P(s)$ is defined as $\sum_{U(p)=s} 2^{-|p|}$. *Chaitin's halting probability* Ω is defined by $\Omega = \sum_{p \in \operatorname{dom} U} 2^{-|p|}$. For any $\alpha \in \mathbb{R}$, we say that α is *weakly Chaitin random* if there exists $c \in \mathbb{N}$ such that $n - c \le H(\alpha\restriction_n)$ for all $n \in \mathbb{N}^+$ [4, 5]. Then [4] showed that Ω is weakly

Chaitin random. For any $\alpha \in \mathbb{R}$, we say that α is *Chaitin random* if $\lim_{n\to\infty} H(\alpha\restriction_n) - n = \infty$ [4, 5]. Obviously, for every $\alpha \in \mathbb{R}$, if α is Chaitin random, then α is weakly Chaitin random. We can show that the converse also hold. Thus, for every $\alpha \in \mathbb{R}$, α is weakly Chaitin random if and only if α is Chaitin random (see Chaitin [5] for the proof and historical detail). Thus Ω is Chaitin random.

2.2. Partial randomness

In the works [10, 11], we generalized the notion of the randomness of a real so that *the degree of the randomness*, which is often referred to as *the partial randomness* recently [2, 8, 3], can be characterized by a real T with $0 \le T \le 1$ as follows.

Definition 2.1 (weak Chaitin T-randomness). *Let $T \ge 0$. For any $\alpha \in \mathbb{R}$, we say that α is weakly Chaitin T-random if there exists $c \in \mathbb{N}$ such that $Tn - c \le H(\alpha\restriction_n)$ for all $n \in \mathbb{N}^+$.* □

Definition 2.2 (T-compressibility). *Let $T \ge 0$. For any $\alpha \in \mathbb{R}$, we say that α is T-compressible if $H(\alpha\restriction_n) \le Tn + o(n)$, which is equivalent to* $\limsup_{n\to\infty} H(\alpha\restriction_n)/n \le T$. □

In the case of $T = 1$, the weak Chaitin T-randomness results in the weak Chaitin randomness. For every $T \in [0, 1]$ and every $\alpha \in \mathbb{R}$, if α is weakly Chaitin T-random and T-compressible, then

$$\lim_{n\to\infty} \frac{H(\alpha\restriction_n)}{n} = T. \tag{3}$$

The left-hand side of (3) is referred to as the *compression rate* of a real α in general. Note, however, that (3) does not necessarily imply that α is weakly Chaitin T-random. Thus, the notion of partial randomness is a stronger representation of the notion of compression rate.

Definition 2.3 (Chaitin T-randomness, Tadaki [10, 11]). *Let $T \ge 0$. For any $\alpha \in \mathbb{R}$, we say that α is Chaitin T-random if* $\lim_{n\to\infty} H(\alpha\restriction_n) - Tn = \infty$. □

In the case of $T = 1$, the Chaitin T-randomness results in the Chaitin randomness. Obviously, for every $T \in [0, 1]$ and every $\alpha \in \mathbb{R}$, if α is Chaitin T-random, then α is weakly Chaitin T-random. However, in 2005 Reimann and Stephan [8] showed that, in the case of $T < 1$, the converse does not necessarily hold.

3. The previous works: Rigorous results

In this section, we summarize the main rigorous results of the statistical mechanical interpretation of AIT, developed by our former works [13, 14]. We first introduce the notion of thermodynamic quantities into AIT in the following manner.

In statistical mechanics, the thermodynamic quantities: free energy $F_{\rm sm}(T)$, energy $E_{\rm sm}(T)$, entropy $S_{\rm sm}(T)$, and specific heat $C_{\rm sm}(T)$ at temperature T are given as follows.

$$\begin{aligned} F_{\rm sm}(T) &= -k_{\rm B}T \ln Z_{\rm sm}(T), & E_{\rm sm}(T) &= \frac{1}{Z_{\rm sm}(T)} \sum_{x\in X} E_x e^{-\frac{E_x}{k_{\rm B}T}}, \\ S_{\rm sm}(T) &= \frac{E_{\rm sm}(T) - F_{\rm sm}(T)}{T}, & C_{\rm sm}(T) &= \frac{d}{dT} E_{\rm sm}(T), \end{aligned} \tag{4}$$

where $Z_{\rm sm}(T)$ is the partition function given by (2).[2] We introduce the notion of thermodynamic quantities into AIT by performing Replacements 1.1 for the thermodynamic quantities (2) and (4) in statistical mechanics.[3] For that purpose, we choose a particular enumeration $q_1, q_2, q_3, q_4, \dots$ of the infinite set $\operatorname{dom} U$ first. Then, motivated by the formulae (2) and (4), and taking into account Replacements 1.1, we introduce the notion of thermodynamic quantities into AIT as follows.

[2] For the thermodynamic quantities in statistical mechanics, see e.g. Chapter 16 of [1] and Chapter 2 of [15].

[3] To be precise, the partition function is not a thermodynamic quantity but a statistical mechanical quantity.

Definition 3.1 (thermodynamic quantities in AIT, [13]). *Let T be any real with $T > 0$.*

(i) *The partition function $Z(T)$ at temperature T is defined as $\lim_{k\to\infty} Z_k(T)$ where $Z_k(T) = \sum_{i=1}^{k} 2^{-\frac{|q_i|}{T}}$.*

(ii) *The free energy $F(T)$ at temperature T is defined as $\lim_{k\to\infty} F_k(T)$ where $F_k(T) = -T\log_2 Z_k(T)$.*

(iii) *The energy $E(T)$ at temperature T is defined as $\lim_{k\to\infty} E_k(T)$ where $E_k(T) = \sum_{i=1}^{k} |q_i|\, 2^{-\frac{|q_i|}{T}}/Z_k(T)$.*

(iv) *The statistical mechanical entropy $S(T)$ at temperature T is defined as $\lim_{k\to\infty} S_k(T)$ where $S_k(T) = \{E_k(T) - F_k(T)\}/T$.*

(v) *The specific heat $C(T)$ at temperature T is defined as $\lim_{k\to\infty} C_k(T)$ where $C_k(T) = E_k'(T)$, the derived function of $E_k(T)$.* □

Note that $Z(T)$ certainly equals to $\Omega(T)$, which is defined by (1).

Theorem 3.2 (properties of $Z(T)$ and $F(T)$, [10, 11, 13]). *Let $T \in \mathbb{R}$.*

(i) *If $0 < T \leq 1$ and T is computable, then each of $Z(T)$ and $F(T)$ converges and is weakly Chaitin T-random and T-compressible.*

(ii) *If $1 < T$, then $Z(T)$ and $F(T)$ diverge to ∞ and $-\infty$, respectively.* □

Theorem 3.3 (properties of $E(T)$ and $S(T)$, [13]). *Let $T \in \mathbb{R}$.*

(i) *If $0 < T < 1$ and T is computable, then each of $E(T)$, $S(T)$, and $C(T)$ converges and is Chaitin T-random and T-compressible.*

(ii) *If $1 \leq T$, then both $E(T)$ and $S(T)$ diverge to ∞. In the case of $T = 1$, $C(T)$ diverges to ∞.*[4] □

The above two theorems show that if T is a computable real with $0 < T < 1$ then the partial randomness (and therefore the compression rate) of the values of all the thermodynamic quantities in Definition 3.1 equals to the temperature T. These theorems also show that the values of all the thermodynamic quantities diverge when the temperature T exceeds 1. This phenomenon might be regarded as some sort of phase transition in statistical mechanics.

In statistical mechanics or thermodynamics, among all thermodynamic quantities one of the most typical thermodynamic quantities is temperature itself. Inspired by this fact in physics and by the above observation that the temperature T equals to the partial randomness of the values of the thermodynamic quantities in the statistical mechanical interpretation of AIT, the following question arises naturally: Can the partial randomness of the temperature T equal to the temperature T itself in the statistical mechanical interpretation of AIT ? This question is rather self-referential. However, we can answer it affirmatively in the following form.

Theorem 3.4 (fixed point theorem by $Z(T)$, [13]). *For every $T \in (0,1)$, if $Z(T)$ is computable, then T is weakly Chaitin T-random and T-compressible.* □

Theorem 3.5 (fixed point theorem by $F(T)$, [14]). *For every $T \in (0,1)$, if $F(T)$ is computable then T is weakly Chaitin T-random and T-compressible.* □

Theorem 3.6 (fixed point theorem by $E(T)$, [14]). *For every $T \in (0,1)$, if $E(T)$ is computable then T is Chaitin T-random and T-compressible.* □

Theorem 3.7 (fixed point theorem by $S(T)$, [14]). *For every $T \in (0,1)$, if $S(T)$ is computable then T is Chaitin T-random and T-compressible.* □

[4] It is still open whether $C(T)$ diverges or not in the case of $T > 1$.

Each of the above four theorems is just a *fixed point theorem on partial randomness*, where the computability of the value of thermodynamic quantity of AIT at temperature T gives a sufficient condition for a real $T \in (0,1)$ to be a *fixed point on partial randomness*, i.e., to satisfy the property that the partial randomness of T equals to T itself, and therefore to satisfy that

$$\lim_{n\to\infty} \frac{H(T{\restriction}_n)}{n} = T. \tag{5}$$

This means that the compression rate of T equals to T itself. Intuitively, we might interpret the meaning of (5) as follows: Consider imaginarily a file of infinite size whose content is

"The compression rate of this file is 0.100111001……"

When this file is compressed, the compression rate of this file actually equals to 0.100111001……, as the content of this file says. This situation is self-referential and forms a fixed point.

4. Total statistical mechanical interpretation of AIT: Physical argument

In what follows, based on a physical argument we develop a total statistical mechanical interpretation of AIT which attains a perfect correspondence to normal statistical mechanics. In consequence, we justify the interpretation of $\Omega(D)$ as a partition function and clarify the statistical mechanical meaning of the thermodynamic quantities of AIT in Definition 3.1. In the work [12], we developed a statistical mechanical interpretation of the noiseless source coding scheme based on an absolutely optimal instantaneous code by identifying a microcanonical ensemble in the scheme. In a similar manner to [12] we develop a total statistical mechanical interpretation of AIT in what follows. This can be possible because the set $\operatorname{dom} U$ is prefix-free and therefore the action of the optimal computer U can be regarded as an instantaneous code which is extended over an infinite set. Note that, in what follows, we do not stick to the mathematical strictness of the argument and we make an argument on the same level of mathematical strictness as statistical mechanics in physics. We start with some reviews of statistical mechanics.

In statistical mechanics we consider a quantum system $\mathcal{S}_{\text{total}}$ which consists in a large number of identical quantum subsystems. Let N be the number of such subsystems. For example, $N \sim 10^{22}$ for $1\,\text{cm}^3$ of a gas at room temperature. We assume here that each quantum subsystem can be distinguishable from others. Thus, we deal with quantum particles which obey *Maxwell-Boltzmann statistics* and not Bose-Einstein statistics or Fermi-Dirac statistics. Under this assumption, we can identify the ith quantum subsystem $\mathcal{S}_i$ for each $i = 1, \dots, N$. In quantum mechanics, any quantum system is described by a quantum state completely. In statistical mechanics, among all quantum states, energy eigenstates are of particular importance. Any energy eigenstate of each subsystem $\mathcal{S}_i$ can be specified by a number $n = 1, 2, 3, \dots$, called a *quantum number*, where the subsystem in the energy eigenstate specified by n has the energy E_n. Then, any energy eigenstate of the system $\mathcal{S}_{\text{total}}$ can be specified by an N-tuple $(n_1, n_2, \dots, n_N)$ of quantum numbers. If the state of the system $\mathcal{S}_{\text{total}}$ is the energy eigenstate specified by $(n_1, n_2, \dots, n_N)$, then the state of each subsystem $\mathcal{S}_i$ is the energy eigenstate specified by n_i and the system $\mathcal{S}_{\text{total}}$ has the energy $E_{n_1} + E_{n_2} + \cdots + E_{n_N}$. Then, the fundamental postulate of statistical mechanics, called *the principle of equal probability*, is stated as follows.

The Principle of Equal Probability: If the energy of the system $\mathcal{S}_{\text{total}}$ is known to have a constant value in the range between E and $E + \delta E$, where δE is the indeterminacy in measurement of the energy of the system $\mathcal{S}_{\text{total}}$, then the system $\mathcal{S}_{\text{total}}$ is equally likely to be in any energy eigenstate specified by $(n_1, n_2, \dots, n_N)$ such that $E \le E_{n_1} + E_{n_2} + \cdots + E_{n_N} \le E + \delta E$. □

Let $\Theta(E, N)$ be the total number of energy eigenstates of $\mathcal{S}_{\text{total}}$ specified by $(n_1, n_2, \dots, n_N)$ such that $E \le E_{n_1} + E_{n_2} + \cdots + E_{n_N} \le E + \delta E$. The above postulate states that any energy eigenstate of $\mathcal{S}_{\text{total}}$ whose energy lies between E and $E + \delta E$ occurs with the probability $1/\Theta(E, N)$. This uniform distribution of energy eigenstates whose energy lie between E and $E + \delta E$ is called a *microcanonical ensemble*. In statistical mechanics, the *entropy* $S(E, N)$ of the system $\mathcal{S}_{\text{total}}$ is then defined by $S(E, N) = k_{\text{B}} \ln \Theta(E, N)$. The average energy ε per one subsystem is given by E/N. In a normal case where ε has a finite value, the entropy $S(E, N)$ is proportional to N. On the other hand, the indeterminacy δE of the energy contributes to $S(E, N)$ through the term $k_{\text{B}} \ln \delta E$, which can be ignored compared to N unless δE is too small. Thus the magnitude of the indeterminacy δE of the energy does not matter to the value of the entropy $S(E, N)$ unless it is too small. The *temperature* $T(E, N)$ of the system $\mathcal{S}_{\text{total}}$ is defined by

$$\frac{1}{T(E, N)} = \frac{\partial S}{\partial E}(E, N).$$

Thus the temperature is a function of E and N.

Now we give a total statistical mechanical interpretation to AIT. As considered in [4], think of the optimal computer U as a decoding equipment at the receiving end of a noiseless binary communication channel. Regard its programs (i.e., finite binary strings in $\operatorname{dom} U$) as codewords and regard the result of the computation by U, which is a finite binary string, as a decoded "symbol." Since $\operatorname{dom} U$ is a prefix-free set, such codewords form what is called an "instantaneous code," so that successive symbols sent through the channel in the form of concatenation of codewords can be separated.

For establishing the total statistical mechanical interpretation of AIT, we assume that the infinite binary string sent through the channel is generated by infinitely repeated tosses of a fair coin. Under this assumption, the infinite binary string is referred to as *the channel infinite string* hereafter. For each $r \in \{0,1\}^*$, let $Q(r)$ be the probability that the channel infinite string has the prefix r. It follows that $Q(r) = 2^{-|r|}$. Thus, the channel infinite string is the random variable drawn according to Lebesgue measure on $\{0,1\}^\infty$. Note that the success probability for U to decode one symbol at the receiving end of the channel infinite string equals to Chaitin's halting probability Ω since $\Omega = \sum_{p \in \operatorname{dom} U} Q(p)$. Since this success probability can be regarded as a probability for U to halt at the receiving end of the channel infinite string, we will establish the total statistical mechanical interpretation of AIT, based on a probability measure which gives Chaitin's Ω the meaning of *halting probability* actually. In addition, note that the probability to get a finite binary string s as the first decoded symbol by U at the receiving end of the channel infinite string equals to $P(s)$ since $P(s) = \sum_{U(p)=s} Q(p)$.

Let N be a large number, say $N \sim 10^{22}$. We relate AIT to the statistical mechanics reviewed above in the following manner. Among all infinite binary strings, consider infinite binary strings of the form $p_1 p_2 \cdots p_N \alpha$ with $p_1, p_2, \dots, p_N \in \operatorname{dom} U$ and $\alpha \in \{0,1\}^\infty$. For each i, the ith slot fed by p_i corresponds to the ith quantum subsystem $\mathcal{S}_i$. On the other hand, the ordered sequence of the 1st slot, the 2nd slot, ..., and the Nth slot corresponds to the quantum system $\mathcal{S}_{\text{total}}$. We relate a codeword $p \in \operatorname{dom} U$ to an energy eigenstate of a quantum subsystem, and relate a codeword length $|p|$ to an energy E_n of the energy eigenstate of the quantum subsystem. Then, a finite binary string $p_1 \cdots p_N$ corresponds to an energy eigenstate of $\mathcal{S}_{\text{total}}$ specified by $(n_1, \dots, n_N)$. Thus, $|p_1 \cdots p_N| = |p_1| + \cdots + |p_N|$ corresponds to the energy $E_{n_1} + \cdots + E_{n_N}$ of the energy eigenstate of $\mathcal{S}_{\text{total}}$.

We define a subset $C(L, N)$ of $\{0,1\}^*$ as the set of all finite binary strings of the form $p_1 \cdots p_N$ with $p_i \in \operatorname{dom} U$ whose total length $|p_1 \cdots p_N|$ lie between L and $L + \delta L$. Then, $\Theta(L, N)$ is defined as the total number of elements of the finite set $C(L, N)$. Therefore, $\Theta(L, N)$ is the total number of all concatenations of N codewords whose total length lie between L and $L + \delta L$. It follows that if $p_1 \cdots p_N \in C(L, N)$, then $2^{-(L+\delta L)} \le Q(p_1 \cdots p_N) \le 2^{-L}$. Thus, all concatenations $p_1 \cdots p_N \in C(L, N)$ of N codewords occur in a prefix of the channel infinite string with the same probability 2^{-L}. Note here that we care nothing about the magnitude of

δL, as in the case of statistical mechanics. Thus, the following principle, called *the principle of equal conditional probability*, holds.

The Principle of Equal Conditional Probability: Given that a concatenation of N codewords of total length L occurs in a prefix of the channel infinite string, all such concatenations occur with the same probability $1/\Theta(L, N)$. □

We introduce a microcanonical ensemble into AIT in this manner. Thus, we can develop a certain sort of statistical mechanics on AIT. Note that, in statistical mechanics, the principle of equal probability is just a conjecture which is not yet proved completely in a realistic physical system. On the other hand, in our total statistical mechanical interpretation of AIT, the principle of equal conditional probability is automatically satisfied.

The *statistical mechanical entropy* $S(L, N)$ is defined by

$$S(L, N) = \log_2 \Theta(L, N). \tag{6}$$

The *temperature* $T(L, N)$ is then defined by

$$\frac{1}{T(L, N)} = \frac{\partial S}{\partial L}(L, N). \tag{7}$$

Thus, the temperature is a function of L and N.

According to the theoretical development of equilibrium statistical mechanics,[5] we can introduce a *canonical ensemble* into AIT in the following manner. We investigate the probability distribution of the left-most codeword p_1 of the channel infinite string, given that a concatenation of N codewords of total length L occurs in a prefix of the channel infinite string. For each $p \in \{0,1\}^*$, let $R(p)$ be the probability that the left-most codeword of the channel infinite string is p, given that a concatenation of N codewords of total length L occurs in a prefix of the channel infinite string. Then, based on the principle of equal conditional probability, we see that $R(p) = \Theta(L - |p|, N-1)/\Theta(L, N)$. From the general definition (6) of statistical mechanical entropy, we thus have

$$R(p) = 2^{S(L-|p|,N-1)-S(L,N)}. \tag{8}$$

Let $E(L, N)$ be the expected length of the left-most codeword of the channel infinite string, given that a concatenation of N codewords of total length L occurs in a prefix of the channel infinite string. Then, the following equality is expected to hold:

$$S(L, N) = S(E(L, N), 1) + S(L - E(L, N), N - 1). \tag{9}$$

Here, the term $S(L, N)$ in the left-hand side denotes the statistical mechanical entropy of the whole concatenation of N codewords of total length L. On the other hand, the first term $S(E(L, N), 1)$ in the right-hand side denotes the statistical mechanical entropy of the left-most codeword of the concatenation of N codewords of total length L while the second term $S(L - E(L, N), N - 1)$ in the right-hand side denotes the statistical mechanical entropy of the remaining $N - 1$ codewords of the concatenation of N codewords of total length L. Thus, the equality (9) represents the additivity of the statistical mechanical entropy. We assume here that the equality (9) holds.

By expanding $S(L - |p|, N - 1)$ around the "equilibrium point" $L - E(L, N)$, we have

$$\begin{aligned} S(L - |p|, N - 1) &= S(L - E(L, N) + E(L, N) - |p|, N - 1) \\ &= S(L - E(L, N), N - 1) + \frac{\partial S}{\partial L}(L - E(L, N), N - 1)(E(L, N) - |p|). \end{aligned} \tag{10}$$

[5] We follow the argument of Section 16-1 of Callen [1] in particular.

Here, we ignore the higher order terms than the first order. Since $N \gg 1$ and $L \gg E(L,N)$, using the definition (7) of temperature we have

$$\frac{\partial S}{\partial L}(L - E(L,N), N-1) = \frac{\partial S}{\partial L}(L,N) = \frac{1}{T(L,N)}. \tag{11}$$

Hence, by (10) and (11), we have

$$S(L - |p|, N-1) = S(L - E(L,N), N-1) + \frac{1}{T(L,N)}(E(L,N) - |p|). \tag{12}$$

Thus, using (8), (9), and (12), we obtain $R(p) = 2^{\frac{E(L,N)-T(L,N)S(E(L,N),1)}{T(L,N)}} 2^{-\frac{|p|}{T(L,N)}}$. Then, according to statistical mechanics or thermodynamics we define the *free energy* $F(L,N)$ of the left-most codeword of the concatenation of N codewords of total length L by

$$F(L,N) = E(L,N) - T(L,N)S(E(L,N),1). \tag{13}$$

It follows that

$$R(p) = 2^{\frac{F(L,N)}{T(L,N)}} 2^{-\frac{|p|}{T(L,N)}}. \tag{14}$$

Using $\sum_{p \in \operatorname{dom} U} R(p) = 1$ we see that, for any $p \in \operatorname{dom} U$,

$$R(p) = \frac{1}{Z(T(L,N))} 2^{-\frac{|p|}{T(L,N)}}, \tag{15}$$

where $Z(T)$ is defined by

$$Z(T) = \sum_{p \in \operatorname{dom} U} 2^{-\frac{|p|}{T}} \qquad (T > 0). \tag{16}$$

$Z(T)$ is called the *partition function* (of the left-most codeword of the channel infinite string). Thus, in our total statistical mechanical interpretation of AIT, the partition function $Z(T)$ has exactly the same form as $\Omega(D)$, which is defined by (1). The distribution in the form of $R(p)$ is called a *canonical ensemble* in statistical mechanics.

Then, using (14) and (15), $F(L,N)$ is calculated as

$$F(L,N) = F(T(L,N)), \tag{17}$$

where $F(T)$ is defined by

$$F(T) = -T \log_2 Z(T) \qquad (T > 0). \tag{18}$$

On the other hand, from the definition of $R(p)$, $E(N,L)$ is calculated as $E(N,L) = \sum_{p \in \operatorname{dom} U} |p| R(p)$. Thus, we have

$$E(L,N) = E(T(L,N)), \tag{19}$$

where $E(T)$ is defined by

$$E(T) = \frac{1}{Z(T)} \sum_{p \in \operatorname{dom} U} |p| 2^{-\frac{|p|}{T}} \qquad (T > 0). \tag{20}$$

Then, using (13), (17), and (19), the statistical mechanical entropy $S(E(L,N),1)$ of the left-most codeword of the concatenation of N codewords of total length L is calculated as $S(E(L,N),1) = S(T(L,N))$, where $S(T)$ is defined by

$$S(T) = \{E(T) - F(T)\}/T \qquad (T > 0). \tag{21}$$

Note that the statistical mechanical entropy $S(E(L,N),1)$ coincides with the *Shannon entropy* $-\sum_{p \in \operatorname{dom} U} R(p) \log_2 R(p)$ of the distribution $R(p)$.

Finally, the *specific heat* $C(T)$ of the left-most codeword of the channel infinite string is defined by

$$C(T) = E'(T) \qquad (T > 0), \tag{22}$$

where $E'(T)$ is the derived function of $E(T)$.

Thus, a total statistical mechanical interpretation of AIT can be established, based on a physical argument. We can check that the formulas in this argument: the partition function (16), the free energy (18), the expected length of the left-most codeword (20), the statistical mechanical entropy (21), and the specific heat (22) correspond to the thermodynamic quantities of AIT in Definition 3.1. Thus, the statistical mechanical meaning of the notion of thermodynamic quantities of AIT in Definition 3.1 is clarified by this argument.

5. Concluding remarks

In this paper, we have developed a total statistical mechanical interpretation of AIT which actualizes a perfect correspondence to normal statistical mechanics. However, the argument used in the development is on the same level of mathematical strictness as normal statistical mechanics in physics. Thus, we try to make the argument a rigorous form in a future study. This effort might stimulate a further unexpected development of the mathematical research of AIT.

Acknowledgments

This work was supported by KAKENHI, Grant-in-Aid for Scientific Research (C) (20540134), by SCOPE of the Ministry of Internal Affairs and Communications of Japan, and by CREST of the Japan Science and Technology Agency.

References

[1] H. B. Callen, *Thermodynamics and an Introduction to Thermostatistics*, 2nd ed. John Wiley & Sons, Inc., Singapore, 1985.

[2] C. S. Calude, L. Staiger, and S. A. Terwijn, "On partial randomness," *Ann. Pure Appl. Logic*, vol. 138, pp. 20–30, 2006.

[3] C. S. Calude and M. A. Stay, "Natural halting probabilities, partial randomness, and zeta functions," *Inform. and Comput.*, vol. 204, pp. 1718–1739, 2006.

[4] G. J. Chaitin, "A theory of program size formally identical to information theory," *J. Assoc. Comput. Mach.*, vol. 22, pp. 329–340, 1975.

[5] G. J. Chaitin, *Algorithmic Information Theory*. Cambridge University Press, Cambridge, 1987.

[6] A. Dembo and O. Zeitouni, *Large Deviations Techniques and Applications*, 2nd ed. Springer, New York, 1998.

[7] F. Reif, *Fundamentals of Statistical and Thermal Physics*. McGraw-Hill, Inc., Singapore, 1965.

[8] J. Reimann and F. Stephan, On hierarchies of randomness tests. Proceedings of the 9th Asian Logic Conference, World Scientific Publishing, August 16-19, 2005, Novosibirsk, Russia.

[9] D. Ruelle, *Statistical Mechanics, Rigorous Results*, 3rd ed. Imperial College Press and World Scientific Publishing Co. Pte. Ltd., Singapore, 1999.

[10] K. Tadaki, Algorithmic information theory and fractal sets. Proceedings of 1999 Workshop on Information-Based Induction Sciences (IBIS'99), pp. 105–110, August 26-27, 1999, Syuzenji, Shizuoka, Japan. In Japanese.

[11] K. Tadaki, "A generalization of Chaitin's halting probability Ω and halting self-similar sets," *Hokkaido Math. J.*, vol. 31, pp. 219–253, 2002.

[12] K. Tadaki, A statistical mechanical interpretation of instantaneous codes. Proceedings of 2007 IEEE International Symposium on Information Theory (ISIT2007), pp. 1906–1910, June 24-29, 2007, Nice, France.

[13] K. Tadaki, A statistical mechanical interpretation of algorithmic information theory. Local Proceedings of Computability in Europe 2008 (CiE 2008), pp. 425–434, June 15-20, 2008, University of Athens, Greece. Electronic Version Available: `http://www.cs.swan.ac.uk/cie08/cie2008-local.pdf`

[14] K. Tadaki, Fixed point theorems on partial randomness. Proceedings of the Symposium on Logical Foundations of Computer Science 2009 (LFCS'09), Lecture Notes in Computer Science, Springer-Verlag, vol. 5407, pp. 422–440, 2009.

[15] M. Toda, R. Kubo, and N. Saitô, *Statistical Physics* I. *Equilibrium Statistical Mechanics*, 2nd ed. Springer, Berlin, 1992.

Mathematical Aspects of Generalized Entropies and their Applications IOP Publishing
Journal of Physics: Conference Series **201** (2010) 012007 doi:10.1088/1742-6596/201/1/012007

Meaning of an escort distribution and τ-transformation

Masaru Tanaka

8-19-1 Nanakuma, Johnan-Ku, Fukuoka 814-0180, JAPAN

E-mail: `sieger@math.sci.fukuoka-u.ac.jp`

Abstract. Tsallis entropy, which is one of nonextensive entropies, gives a q-normal distribution as an equilibrium probability density function. Although a q-normal distribution is popular, there exists a problem what it means by calculating an expectation value with a corresponding escort distribution not a q-normal distribution itself. But we have an amazing property such that an escort distribution obtained by a q-normal distribution with a parameter q and a variance is another q-normal distribution with a different value of q and a scaled variance. Therefore calculating an expectation value with an escort distribution corresponds to calculating the expectation value with another q-normal distribution, that is to say, an escort distribution is nothing but another q-normal distribution. However it still remains the question why an expectation value should be calculated by another q-normal distribution. We call the procedure to get another q-normal distribution from a q-normal distribution through an escort distribution τ-transformation. This τ-transformation keeps a support of a q-normal distribution invariant, and makes the tails of a q-normal distribution thicker/thinner depending on a value of q. Thus a τ-transformation looks like a kind of multiresolutional analysis, if we consider a q-normal distribution as a window function.

1. Introduction

Tsallis entropy[1], which is one of nonextensive entropies, gives a q-normal distribution as an equilibrium probability density function. The q-normal distribution plays a lot of roles in various fields, for instance, as Cauchy distribution, t-distribution and Gaussian distribution with suitable values of q in statistics[1][2], as a representation of a projected rotation group in image recognition[3], as an instanton in a moduli space of a nonlinear σ model with Fisher metric in elementary particle physics[4], as a Barenblatt-Pattle self-similar solution of a porous medium equation in condensed matter physics[1][5], and so on.

Although a q-normal distribution is popular as above, there exists a problem what it means by calculating an expectation value with a corresponding escort distribution not a q-normal distribution itself. For this problem, one of possible answers is to consider an escort distribution as an actual probability density function, and a q-normal distribution as just an element of a suitable linear space[6]. Such a nonlinear mapping gives some mathematical meanings to a relation between a corresponding escort distribution and a q-normal distribution, but it can't answer why we consider such a nonlinear mapping. However there is a wonderful property such that an escort distribution obtained by a q-normal distribution with $q = a$ and a variance σ^2 is equivalent to another q-normal distribution with $q = 2 - \frac{1}{a}$ and a variance $\frac{3-a}{1+a}\sigma^2$.

Then calculating an expectation value with an escort distribution corresponds to calculating the expectation value with another associated q-normal distribution, but it still remains the question why an expectation value should be calculated by another q-normal distribution. We call the procedure to get another q-normal distribution from a given q-normal distribution through an escort distribution τ-transformation. This τ-transformation keeps a support of a q-normal distribution invariant, and makes the tails of a q-normal distribution thicker/thinner depending on a value of q. Therefore, a τ-transformation looks like a kind of multiresolutional analysis, if we consider a q-normal distribution as a window function.

In this paper, we consider this τ-transformation and a sequence of q-normal distributions obtained by successive τ-transformations. Then the meaning of an escort distribution is elucidated. That is to say, an escort distribution is nothing but another q-normal distribution. It is shown that the sequence of q-normal distributions obtained by successive τ-transformations gives a rule to change a resolution in observing a phenomenon following some probability, from coarse to fine or in reverse. We also show that the number of possible successive τ-transformations is limited depending on the value of q.

2. q-normal distribution

There are mainly two ways of getting a q-normal distribution: one is from the maximum entropy principle with Tsallis entropy[1][2], the other is from an extension of Gaussian law of error[7]. Of course, these two methods give the same q-normal distribution $p_q(x|\mu,\sigma)$ where μ and σ are a mean and a standard deviation, respectively:

$$p_q(x|\mu,\sigma) = \frac{1}{Z_{q,\sigma}}\left(\left[1-\frac{1-q}{3-q}\frac{(x-\mu)^2}{\sigma^2}\right]_+\right)^{\frac{1}{1-q}} \tag{1}$$

where

$$[u]_+ = \max.(0,u) \ , \tag{2}$$

and

$$Z_{q,\sigma} = \begin{cases} \sqrt{\frac{3-q}{q-1}\pi}\,\frac{\Gamma\left(\frac{1}{q-1}-\frac{1}{2}\right)}{\Gamma\left(\frac{1}{q-1}\right)}\sigma & \text{for } 1 \le q < 3 \\[2ex] \sqrt{\frac{3-q}{1-q}\pi}\,\frac{\Gamma\left(\frac{2-q}{1-q}\right)}{\Gamma\left(\frac{2-q}{1-q}+\frac{1}{2}\right)}\sigma & \text{for } q < 1 \end{cases} \tag{3}$$

The limitation of a range of q comes up so that the normalization factor $Z_{q,\sigma}$, the mean μ and the variance σ^2 must be finite in calculating their expectations with an associated escort distribution:

$$P_q(x|\mu,\sigma) = \frac{1}{\Omega_{q,\sigma}}\,p_q(x|\mu,\sigma)^q \ , \tag{4}$$

where $\Omega_{q,\sigma}$ is a normalization factor. The capital letter P stands for an escort distribution and the small letter p is used for a q-normal distribution.

The normalization factor $\Omega_{q,\sigma}$ is related to the normalization factor $Z_{q,\sigma}$ as

$$\Omega_{q,\sigma} = \frac{3-q}{2}\left(Z_{q,\sigma}\right)^{q-1} \ . \tag{5}$$

Because we have, for the case of $1 \le q < 3$ with the eq.(A.1) for $\alpha = 0, \beta = \frac{q}{q-1}, \lambda = 2$,

$$\int_{-\infty}^{\infty}\left\{1-\frac{1-q}{3-q}\frac{(x-\mu)^2}{\sigma^2}\right\}^{\frac{q}{1-q}}dx = \sqrt{\frac{3-q}{q-1}}\,\sigma\int_{-\infty}^{\infty}\frac{1}{(1+y^2)^{\frac{q}{q-1}}}dy$$

$$= 2\sqrt{\frac{3-q}{q-1}}\sigma\int_0^\infty \frac{1}{(1+y^2)^{\frac{q}{q-1}}}dy = 2\sqrt{\frac{3-q}{q-1}}\sigma\frac{1}{2}\frac{\Gamma\left(\frac{q}{q-1}-\frac{1}{2}\right)\Gamma\left(\frac{1}{2}\right)}{\Gamma\left(\frac{q}{q-1}\right)}$$

$$= \frac{3-q}{2}\sqrt{\frac{3-q}{q-1}\pi}\frac{\Gamma\left(\frac{1}{q-1}-\frac{1}{2}\right)}{\Gamma\left(\frac{1}{q-1}\right)}\sigma = \frac{3-q}{2}Z_{q,\sigma}\ , \tag{6}$$

where $y=\sqrt{\frac{q-1}{3-q}}\frac{x-\mu}{\sigma}$, and for the case of $q<1$ with the eq.(A.2) for $\alpha=0, \beta=\frac{q}{1-q}, \lambda=2$,

$$\int_{\Lambda_-}^{\Lambda_+}\left\{1-\frac{1-q}{3-q}\frac{(x-\mu)^2}{\sigma^2}\right\}^{\frac{q}{1-q}}dx = \sqrt{\frac{3-q}{1-q}}\sigma\int_{-1}^{1}\frac{1}{(1-y^2)^{\frac{q}{1-q}}}dy$$

$$= 2\sqrt{\frac{3-q}{1-q}}\sigma\int_0^1\frac{1}{(1-y^2)^{\frac{q}{1-q}}}dy = 2\sqrt{\frac{3-q}{1-q}}\sigma\frac{1}{2}\frac{\Gamma\left(\frac{1}{2}\right)\Gamma\left(\frac{q}{1-q}+1\right)}{\Gamma\left(\frac{1}{2}+\frac{q}{1-q}+1\right)}$$

$$= \sqrt{\frac{3-q}{1-q}\pi}\frac{\Gamma\left(\frac{1}{1-q}\right)}{\Gamma\left(\frac{1}{2}+\frac{1}{1-q}\right)}\sigma = \frac{3-q}{2}\sqrt{\frac{3-q}{1-q}\pi}\frac{\Gamma\left(\frac{2-q}{1-q}\right)}{\Gamma\left(\frac{1}{2}+\frac{2-q}{1-q}\right)}\sigma = \frac{3-q}{2}Z_{q,\sigma}\ , \tag{7}$$

where $y=\sqrt{\frac{1-q}{3-q}}\frac{x-\mu}{\sigma}$, and the cut-offs $\Lambda_\pm=\mu\pm\sqrt{\frac{3-q}{1-q}}\sigma$ appear for $\left[1-\frac{1-q}{3-q}\frac{(x-\mu)^2}{\sigma^2}\right]_+$.

Therefore, an escort distribution associated with a q-normal distribution can be written in

$$P_q(x|\mu,\sigma) = \frac{2}{3-q}\frac{1}{Z_{q,\sigma}}\left(\left[1-\frac{1-q}{3-q}\frac{(x-\mu)^2}{\sigma^2}\right]_+\right)^{\frac{q}{1-q}} = \frac{2}{3-q}\frac{1}{(Z_{q,\sigma})^{1-q}}p_q(x|\mu,\sigma)^q\ . \tag{8}$$

From now on, we omit the symbol $[\bullet]_+$ for the sake of simplicity.

Although a q-normal distribution is one of the probability density functions, in Tsallis statistics, the expectation values must be evaluated by its associated escort distribution not the q-normal distribution itself. We don't know why this is required yet, but in the future it must be answered, if Tsallis statistics has some meanings in science.

3. τ-transformation

Let's begin with observing a following transformation on a given q-normal distribution:

$$q' = 2-\frac{1}{q}\ ,\ \sigma'^2 = \frac{3-q}{1+q}\sigma^2\ . \tag{9}$$

Under this transformation, because of

$$\frac{1-q'}{3-q'} = \frac{1-\left(2-\frac{1}{q}\right)}{3-\left(2-\frac{1}{q}\right)} = \frac{\frac{1}{q}-1}{\frac{1}{q}+1} = \frac{1-q}{1+q}\ , \tag{10}$$

we have

$$\frac{1-q'}{3-q'}\frac{1}{\sigma'^2} = \frac{1-q}{1+q}\frac{1}{\frac{3-q}{1+q}\sigma^2} = \frac{1-q}{3-q}\frac{1}{\sigma^2}\ . \tag{11}$$

Furthermore, the following relation holds

$$\frac{1}{1-q'} = \frac{1}{1-\left(2-\frac{1}{q}\right)} = \frac{1}{\frac{1}{q}-1} = \frac{q}{1-q}\ . \tag{12}$$

Therefore we have

$$\left\{1-\frac{1-q'}{3-q'}\frac{(x-\mu)^2}{\sigma'^2}\right\}^{\frac{1}{1-q'}}=\left\{1-\frac{1-q}{3-q}\frac{(x-\mu)^2}{\sigma^2}\right\}^{\frac{q}{1-q}}. \tag{13}$$

Amazingly, gamma functions change under this transformation as follows, for $1 \le q' < 3$,

$$\begin{aligned}\frac{\Gamma\left(\frac{1}{q'-1}-\frac{1}{2}\right)}{\Gamma\left(\frac{1}{q'-1}\right)} &= \frac{\Gamma\left(\frac{q}{q-1}-\frac{1}{2}\right)}{\Gamma\left(\frac{q}{q-1}\right)}=\frac{\Gamma\left(\frac{1}{q-1}-\frac{1}{2}+1\right)}{\Gamma\left(\frac{1}{q-1}+1\right)} \\ &= \frac{\left(\frac{1}{q-1}-\frac{1}{2}\right)\Gamma\left(\frac{1}{q-1}-\frac{1}{2}\right)}{\frac{1}{q-1}\Gamma\left(\frac{1}{q-1}\right)}=\frac{3-q}{2}\frac{\Gamma\left(\frac{1}{q-1}-\frac{1}{2}\right)}{\Gamma\left(\frac{1}{q-1}\right)}, \end{aligned} \tag{14}$$

for $q' < 1$,

$$\begin{aligned}\frac{\Gamma\left(\frac{2-q'}{1-q'}\right)}{\Gamma\left(\frac{2-q'}{1-q'}+\frac{1}{2}\right)} &= \frac{\Gamma\left(\frac{1}{1-q}\right)}{\Gamma\left(\frac{1}{1-q}+\frac{1}{2}\right)}=\frac{(1-q)\frac{1}{1-q}\Gamma\left(\frac{1}{1-q}\right)}{\left(\frac{1}{1-q}+\frac{1}{2}\right)^{-1}\left(\frac{1}{1-q}+\frac{1}{2}\right)\Gamma\left(\frac{1}{1-q}+\frac{1}{2}\right)} \\ &= \frac{3-q}{2}\frac{\Gamma\left(\frac{1}{1-q}+1\right)}{\Gamma\left(\frac{1}{1-q}+\frac{1}{2}+1\right)}=\frac{3-q}{2}\frac{\Gamma\left(\frac{2-q}{1-q}\right)}{\Gamma\left(\frac{2-q}{1-q}+\frac{1}{2}\right)}. \end{aligned} \tag{15}$$

Then we have the following relation under the parameter q and the variance transformations,

$$p_{q'}(x|\mu,\sigma')=\frac{2}{3-q}\frac{1}{Z_{q,\sigma}}\left\{1-\frac{1-q}{3-q}\frac{(x-\mu)^2}{\sigma^2}\right\}^{\frac{q}{1-q}}=P_q(x|\mu,\sigma). \tag{16}$$

This relation between a q-normal distribution and an escort distribution obtained by another q-normal distribution enables us to say that *the escort distribution is just another q-normal distribution.* To evaluate an expectation value with an escort distribution means to evaluate the expectation value with another q-normal distribution with a parameter q' and a scaled variance σ'^2. So this is the meaning of an escort distribution.

A procedure to get another q-normal distribution from a given q-normal distribution through an escort distribution is called τ-transformation. τ-transformation keeps a support of a q-normal distribution invariant, and makes the tails of a q-normal distribution thicker/thinner depending on q.

If we define

$$m_q(x|\mu,\sigma)=1-\frac{1-q}{3-q}\frac{(x-\mu)^2}{\sigma^2}, \tag{17}$$

a q-normal distribution is written as

$$p_q(x|\mu,\sigma)=\frac{1}{Z_{q,\sigma}}m_q(x|\mu,\sigma)^{\frac{1}{1-q}}. \tag{18}$$

Then its corresponding escort distribution is

$$P_q(x|\mu,\sigma)=\frac{2}{3-q}\frac{1}{Z_{q,\sigma}}m_q(x|\mu,\sigma)^{\frac{q}{1-q}}. \tag{19}$$

Journal of Physics: Conference Series **201** (2010) 012007 doi:10.1088/1742-6596/201/1/012007

Next we define q_n for a τ-transformation:

$$q_n = \frac{q - n\,(1-q)}{1 - n\,(1-q)} \; . \tag{20}$$

Then it satisfies the following relation:

$$\frac{1}{1 - q_{n+1}} = \frac{q_n}{1 - q_n} \; . \tag{21}$$

If a standard deviation σ changes like

$$\sigma_n = \sigma \sqrt{\frac{3-q}{3 - q - 2n\,(1-q)}} \; , \tag{22}$$

we have the following relation

$$m_{q_n}(x|\mu, \sigma_n) = m_q(x|\mu, \sigma) \; . \tag{23}$$

This leads to a property of τ-transformation keeping a support of a q-normal distribution invariant.

The normalization factors $\{Z_{q_n,\sigma_n}\}$ satisfy the following recursion relation

$$Z_{q_{n+1},\sigma_{n+1}} = \frac{3 - q_n}{2} Z_{q_n,\sigma_n} \; . \tag{24}$$

Therefore we have a sequence giving another q-normal distribution through an escort distribution:

$$\begin{aligned} P_{q_n}(x|\mu, \sigma_n) &= \frac{2}{3 - q_n} \frac{1}{Z_{q_n,\sigma_n}} m_{q_n}(x|\mu, \sigma_n)^{\frac{q_n}{1-q_n}} \\ &= \frac{1}{Z_{q_{n+1},\sigma_{n+1}}} m_{q_{n+1}}(x|\mu, \sigma_{n+1})^{\frac{1}{1-q_{n+1}}} = p_{q_{n+1}}(x|\mu, \sigma_{n+1}) \; . \end{aligned} \tag{25}$$

With the eq.(25), we define τ-transformation from a given q-normal distribution $p_{q_n}(x|\mu, \sigma_n)$ to another q-normal distribution $p_{q_{n+1}}(x|\mu, \sigma_{n+1})$ through an escort distribution $P_{q_n}(x|\mu, \sigma_n)$ of $p_{q_n}(x|\mu, \sigma_n)$ as

$$\tau(p_{q_n}(x|\mu, \sigma_n)) = p_{q_{n+1}}(x|\mu, \sigma_{n+1}) \; . \tag{26}$$

Then the number of possible successive τ-transformations is given as follows,

$$\begin{cases} -\infty < n < \dfrac{q}{1-q} & \text{for } 0 < q < 1 \; , \\ -\infty < n < \infty & \text{for } q = 1 \; , \\ -\dfrac{1}{2}\dfrac{5-3q}{q-1} < n < \infty & \text{for } 1 < q < \dfrac{5}{3} \; . \end{cases} \tag{27}$$

The upper bound of a parameter q, that is $\frac{5}{3}$, comes up so that the second moment for a q-normal distribution, which is calculated with the q-normal distribution itself, is finite. The lower bound of a parameter q, that is 0, comes up because the convexity of Tsallis entropy must be conserved.

As the τ-transformation can be applied to every q-normal distribution with a parameter q_n and a variance σ_n^2, we have a sequence such as

$$\{\ldots, p_{q_{n-1}}(x|\mu, \sigma_{n-1})\,, p_{q_n}(x|\mu, \sigma_n)\,, p_{q_{n+1}}(x|\mu, \sigma_{n+1})\,, \ldots\} \; . \tag{28}$$

We call this sequence a τ-sequence (see figure 1 and figure 2). From the point of view of pattern recognition, the τ-sequence gives a rule to change a resolution in observing a phenomenon following some probability, from coarse to fine or in reverse.

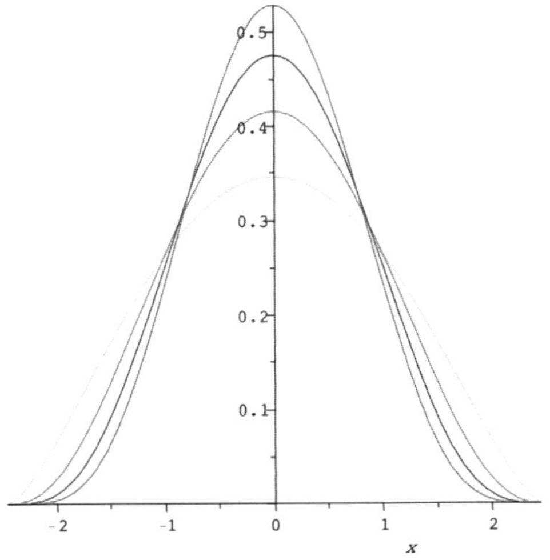

Figure 1. For $q = 0.6$, as n is larger, the standard deviation and the tails of a q-normal distribution are larger and thinner respectively. The support of the q-normal distribution is compact.

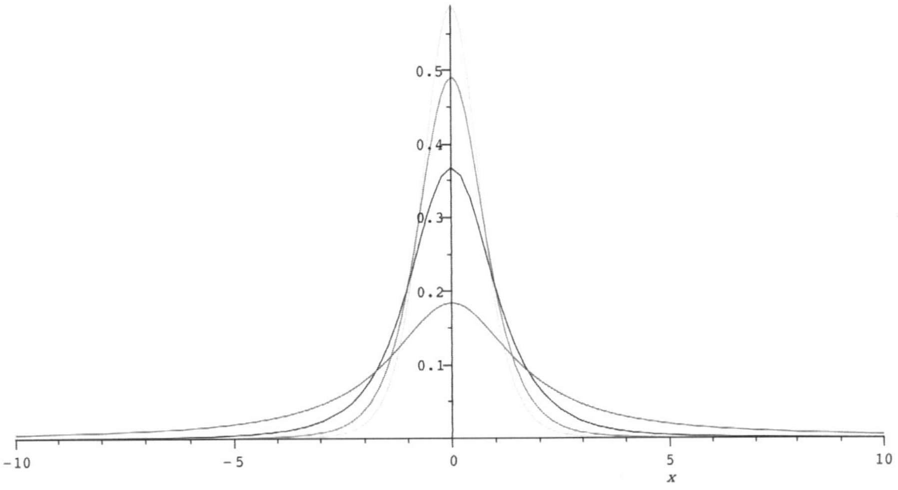

Figure 2. For $q = 1.5$, as n is larger, the standard deviation and the tails of a q-normal distribution are less and thicker respectively. The support of the q-normal distribution is noncompact.

4. δ-transformation

In this section, we consider another transformation which is called δ-transformation. First, let Q_n be

$$Q_n = q - n(1-q) \ . \tag{29}$$

Then it satisfies the following relations:

$$\frac{Q_n}{1-q} = \frac{q_n}{1-q_n} = \frac{1}{1-q_{n+1}} \ . \tag{30}$$

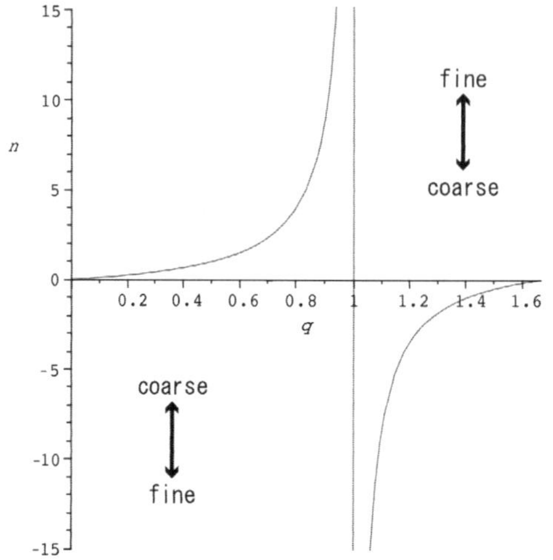

Figure 3. The range of n changes depending on the value of q. For the case of $0 < q < 1$, the less n is, the finer a resolution is. On the other hand, for the case of $1 < q < \frac{5}{3}$, the larger n is, the finer a resolution is. For the case of $q = 1$, a resolution is independent of n.

So the following equations hold

$$Z_{q_{n+1},\sigma_{n+1}} p_{q_{n+1}}(x|\mu,\sigma_{n+1}) = \left(Z_{q_n,\sigma_n}\right)^{q_n} p_{q_n}(x|\mu,\sigma_n)^{q_n} = \left(Z_{q,\sigma}\right)^{Q_n} p_q(x|\mu,\sigma)^{Q_n} \tag{31}$$

If we consider a statistical manifold of q-normal distributions, the basis of tangent space in a suitable coordinate system is

$$\partial_i p_q(x|\mu,\sigma) = \left(\frac{\partial_i p_q(x|\mu,\sigma)^{1-q}}{1-q}\right) p_q(x|\mu,\sigma)^q \ , \tag{32}$$

where ∂_i stands for a partial derivative with respect to some coordinate i.

Then, for q_n and σ_n after some τ-transformations, we have

$$\partial_i p_{q_n}(x|\mu,\sigma_n) = \left(\frac{\partial_i p_{q_n}(x|\mu,\sigma_n)^{1-q_n}}{1-q_n}\right) \frac{\left(Z_{q,\sigma}\right)^{Q_n}}{\left(Z_{q_n,\sigma_n}\right)^{q_n}} p_q(x|\mu,\sigma)^{Q_n} \ . \tag{33}$$

We define δ-transformation as selecting $p_q(x|\mu,\sigma)^{Q_n}$ from the basis, that is

$$\delta(p_{q_n}(x|\mu,\sigma_n)) = p_q(x|\mu,\sigma)^{Q_n} \ . \tag{34}$$

This δ-transformation is used to show that the q-Gaussian family is an invariant manifold for the porous medium equations and the nonlinear Fokker-Planck equations[5].

As the δ-transformation can be applied to every q-normal distribution with a parameter q_n and a variance σ_n^2, we have a sequence such as

$$\left\{\ldots, p_q(x|\mu,\sigma)^{Q_{n-1}}, p_q(x|\mu,\sigma)^{Q_n}, p_q(x|\mu,\sigma)^{Q_{n+1}}, \ldots\right\} \ . \tag{35}$$

This sequence is called a δ-sequence.

5. Conclusions

In this paper, we have given the meaning of an escort distribution obtained by a given q-normal distribution, that is to say, an escort distribution is just another q-normal distribution. We have also given a τ-sequence and a δ-sequence. Then the τ-sequence obtained by successive τ-transformations gives a rule to change a resolution in observing a phenomenon following some probability, from coarse to fine or in reverse. The n-th τ-transformation is characterized by

$$q_n = \frac{q - n(1-q)}{1 - n(1-q)} \;, \; \sigma_n^2 = \frac{3-q}{3-q-2n(1-q)}\sigma^2 \,. \tag{36}$$

Figure 4 shows a part of the τ-sequence and the δ-sequence. They are related through exponentiations $\{q_n\}$ each other.

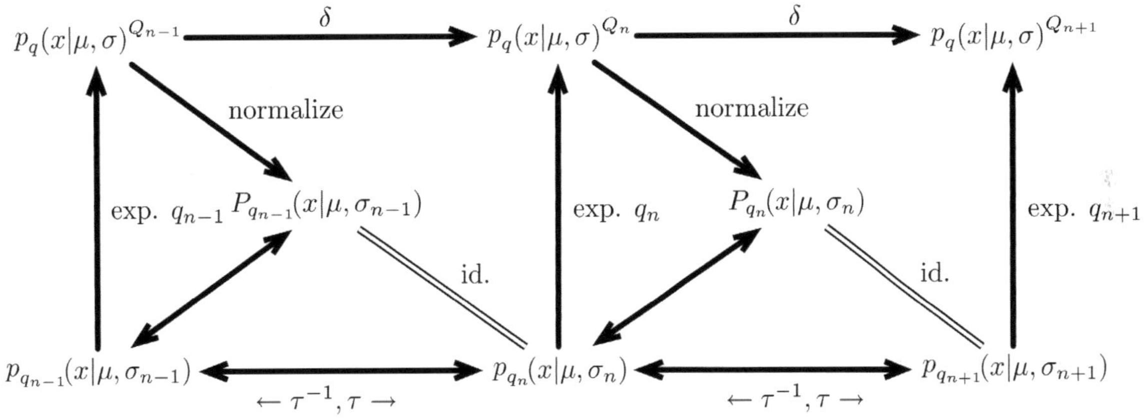

Figure 4. A part of the τ-sequence and the δ-sequence

The meaning of an escort distribution is clear, however it still remains the question why an escort distribution is required.

Finally, this research was partially supported by the Ministry of Education, Science, Sports and Culture, Grant-in-Aid for Exploratory Research, 19654013, 2009.

Appendix A. Integral formula

The following integral formulae can be derived through the definition of a beta function directly. They are quite useful for calculating normalization factors of a q-normal distribution and an escort distribution.

$$\int_0^\infty \frac{dx}{x^\alpha (1+x^\lambda)^\beta} = \frac{1}{\lambda}\frac{\Gamma\left(\beta - \frac{1-\alpha}{\lambda}\right)\Gamma\left(\frac{1-\alpha}{\lambda}\right)}{\Gamma(\beta)} \,, \tag{A.1}$$

where

$$\begin{cases} \alpha < 1 \\ \lambda\,,\,\beta > 0 \\ \lambda\beta > 1-\alpha \end{cases}$$

$$\int_0^1 x^\alpha \left(1 - x^\lambda\right)^\beta dx = \frac{1}{\lambda}\frac{\Gamma\left(\frac{\alpha+1}{\lambda}\right)\Gamma(\beta+1)}{\Gamma\left(\frac{\alpha+1}{\lambda}+\beta+1\right)} \,, \tag{A.2}$$

where

$$\begin{cases} \mathcal{R}e\,[\alpha] & > & -1 \\ \mathcal{R}e\,[\beta] & > & -1 \\ \lambda & > & 0 \end{cases}$$

References

[1] Tsallis C 2009 *Introduction to Nonextensive Statistical Mechanics: Approaching a Complex World*, Springer
[2] Tanaka M 2002 *J. Inst. Elect. Inf. Comm. Eng. Japan* **J85-E-II** 161
[3] Tanaka M 1993 *Pattern Recognition Letters* **14**(8) 671
[4] Yahikozawa S 2004 *Phys. Rev.* E **69** 026122
[5] Ohara A and T. Wada T 2008 *arXiv:0810.0624*
[6] Tanaka M and Hiraoka K 2007 *The 54th symposium on geometry* 152
[7] Suyari H and Tsukada M 2005 *IEEE trans. Inform. Theory* **51** 753
[8] Amari S and Nagaoka H 2000 *Methods of information geometry*, Translations of mathematical monographs **191** American Mathematical Society & Oxford University Press

Mathematical Aspects of Generalized Entropies and their Applications IOP Publishing
Journal of Physics: Conference Series **201** (2010) 012008 doi:10.1088/1742-6596/201/1/012008

q-Gaussian distributions and multiplicative stochastic processes for analysis of multiple financial time series

Aki-Hiro Sato

[1]Department of Applied Mathematics and Physics,
Graduate School of Informatics, Kyoto University,
Yoshida-Honmachi, Sakyo-ku, 606-8501, Kyoto, JAPAN

E-mail: aki@i.kyoto-u.ac.jp

Abstract. This study considers q-Gaussian distributions and stochastic differential equations with both multiplicative and additive noises. In the M-dimensional case a q-Gaussian distribution can be theoretically derived as a stationary probability distribution of the multiplicative stochastic differential equation with both mutually independent multiplicative and additive noises. By using the proposed stochastic differential equation a method to evaluate a default probability under a given risk buffer is proposed.

1. Introduction

Values of risky assets such as stocks, bonds, and loans sometimes unpredictably fluctuate and are not mutually independent of one another. Since financial markets are complex systems consisting of many interacting elements, it is almost undesirable to completely model price fluctuations from microscopic point of view. One of the most prominent approaches seems to manage situations based on characteristics of fluctuations in complex systems, such as power-law distributions [1, 2].

Recently regarding the fat-tailness of distributions of price changes (extremal events) the risk management for risky assets based on financial time series is becoming a crucial issue for financial institutions [3]. The risk management is divided into two parts: an estimation of risks of assets and an assignment of their amount (portfolio) within a limited monetary buffer.

Under the assumption of Gaussianity one sometimes has a tendency to underestimate risks which one is potentially facing. Therefore it is important to establish methodology to evaluate risks of own portfolio taking into account their power-law tailness [4].

Recently the framework of Tsallis statistics has been proposed in order to analyze the fluctuations which come from the complex systems. It is well-known that maximization of Tsallis entropy under constraints on normalization and variance for its escort distribution leads to a q-Gaussian distribution, and that the q-Gaussian distribution has power-law tails when $q > 1$. On the other hand various types of dynamical models of which probability has power law tails. If one can combine statistics and dynamics for analyzing the financial time series with actual data, then one expects to improve its estimation in risk assessment.

In this article the q-Gaussian distribution and a stochastic differential equation with both multiplicative and additive noises are considered in order to analyze multiple time series of a financial market.

Journal of Physics: Conference Series **201** (2010) 012008 doi:10.1088/1742-6596/201/1/012008

This article is organized as follows. In Sec. 2 a brief explanation of Tsallis entropy is shown. In Sec. 3 a stochastic differential equation with both multiplicative and additive noises generating q-Gaussian fluctuations is introduced. In Sec. 4 a portfolio optimization technique under q-Gaussian fluctuations is proposed. In Sec. 5 results of both numerical simulations and empirical calculations are shown. Sec. 6 is devoted to concluding remarks.

2. q-Gaussian on Tsallis statistics

q-Gaussian distributions are successively studied on the context of complex systems or Tsallis statistics. Tsallis entropy [5] is a generalization of entropy and its maximization

$$S_q[p] = \frac{1 - \int_{-\infty}^{\infty} p(x)^q dx}{q-1}, \quad q > 1 \tag{1}$$

under conditions,

$$\int_{-\infty}^{\infty} p(x)dx = 1, \tag{2}$$

$$\int_{-\infty}^{\infty} p^{(q)}(x)x^2 dx = \sigma_q^2, \tag{3}$$

where $p^{(q)}(x) = p^q(x)/\int p^q(x)dx$ is called an escort probability, yields a q-Gaussian distribution

$$p(x) = \frac{1}{Z_q}\Big(1 + \frac{q-1}{(3-q)\sigma_q^2}x^2\Big)^{\frac{1}{1-q}}, \tag{4}$$

$$Z_q^{-1} = \Big(\frac{q-1}{(3-q)\sigma_q^2}\Big)^{-\frac{1}{2}} \frac{\Gamma(\frac{1}{2})\Gamma(\frac{1}{q-1}-\frac{1}{2})}{\Gamma(\frac{1}{q-1})}, \tag{5}$$

where $\Gamma(x)$ is defined as

$$\Gamma(x) = \int_0^{\infty} t^{x-1}e^{-t}dt \quad (x > 0). \tag{6}$$

3. Random multiplicative Processes

It is well known that several types of stochastic differential equations can generate q-Gaussian fluctuations [8, 9, 10, 11]. Nonlinear stochastic differential equations [8], Markov switching model [9, 12], superstatistics of Gaussian processes with a stochastic temperature sampled from a χ^2-distribution [10], and a linear stochastic differential equation with both multiplicative and additive noises [7, 11] have been extensively studied in the context of power law fluctuations.

A stochastic differential equation with both multiplicative and additive noises [11]

$$\frac{dx}{dt} = [\nu + \xi(t)]x(t) + \eta(t), \tag{7}$$

where $\xi(t)$ and $\eta(t)$ represent Gaussian white noises having

$$\langle \xi(t) \rangle = 0, \qquad \langle \xi(t_1)\xi(t_2) \rangle = 2D_\xi \delta(t_1 - t_2), \tag{8}$$

$$\langle \eta(t) \rangle = 0, \qquad \langle \eta(t_1)\eta(t_2) \rangle = 2D_\eta \delta(t_1 - t_2), \tag{9}$$

$$\langle \xi(t)\eta(t) \rangle = 0, \tag{10}$$

yields the Ito/Stratonovich Fokker-Planck equation

$$\frac{\partial P}{\partial t} = -\frac{\partial}{\partial x}\Big[\nu x P\Big] + \frac{\partial^2}{\partial x^2}(D_\xi x^2 + D_\eta)P, \tag{11}$$

$$\frac{\partial P}{\partial t} = -\frac{\partial}{\partial x}\Big[(\nu + D_\xi)xP\Big] + \frac{\partial^2}{\partial x^2}(D_\xi x^2 + D_\eta)P. \tag{12}$$

From Eqs. (11) and (12) under the assumption that $\nu < 0$, $D_\xi > 0$, and $D_\eta > 0$, one obtains a q-Gaussian stationary distribution

$$P_{eq}(x) = \frac{1}{Z}\Big(1 + \frac{D_\xi}{D_\eta}x^2\Big)^{-(\alpha+1)/2}, \tag{13}$$

$$Z^{-1} = \Big(\frac{D_\eta}{D_\xi}\Big)^{\frac{1}{2}}\frac{\Gamma(\frac{1}{2})\Gamma(\frac{\alpha}{2})}{\Gamma(\frac{\alpha+1}{2})}, \tag{14}$$

where Z represents a partition function, and α denotes a characteristic exponent given by $\alpha = -2\nu/D_\xi + 1$ (Ito calculus) and $\alpha = -2\nu/D_\xi$ (Stratonovich calculus).

Furthermore the M-dimensional version of Eq. (7) may be describe as

$$\frac{dx_i}{dt} = [\nu + \xi(t)]\sum_{j=1}^{M} A_{ij}x_j(t) + \eta_i(t), \tag{15}$$

where A_{ij} is a semi-defined matrix, $\eta_i(t)$ represent mutually independent Gaussian white noises,

$$\langle \xi(t)\rangle = 0, \qquad \langle \xi(t_1)\xi(t_2)\rangle = 2D_\xi\delta(t_1 - t_2), \tag{16}$$

$$\langle \eta_i(t)\rangle = 0, \qquad \langle \eta_i(t_1)\eta_j(t_2)\rangle = 2D_\eta\delta_{i,j}\delta(t_1 - t_2), \tag{17}$$

$$\langle \eta_i(t)\xi(t)\rangle = 0. \tag{18}$$

One can find a unitary matrix Q_{ij} which orthogonalizes A_{ij} by means of linear transformation $y_i = \sum_{j=1}^{M} Q_{ij}x_j$ and obtain its eigenvalues λ_i $(i = 1, \ldots, M)$. Applying the linear transformation to Eq. (15) yields

$$\frac{dy_i}{dt} = [\nu + \xi(t)]\lambda_i y_i(t) + \eta'_i(t), \tag{19}$$

where and $\eta'_i(t) = \sum_{j=1}^{M} Q_{ij}\eta_j(t)$. According to properties of white Gaussian noises with homogeneous noise strength and a unitary matrix $\eta'_i(t)$ are also white Gaussian noises. Therefore one obtains a stationary solution from the Ito/Stratonovich Fokker-Planck equation

$$\frac{\partial P_i}{\partial t} = -\frac{\partial}{\partial y_i}\Big[\nu\lambda_i y_i P_i\Big] + \frac{\partial^2}{\partial y_i^2}(\lambda_i^2 D_\xi y_i^2 + D_\eta)P_i, \tag{20}$$

$$\frac{\partial P_i}{\partial t} = -\frac{\partial}{\partial y_i}\Big[(\nu\lambda_i + D_\xi\lambda_i^2)y_i P_i\Big] + \frac{\partial^2}{\partial y_i^2}(\lambda_i^2 D_\xi y_i^2 + D_\eta)P_i, \tag{21}$$

corresponding to Eq. (19)

$$P_j(y_j) = \frac{1}{Z_j}\Big(1 + \lambda_j^2\frac{D_\xi}{D_\eta}y_j^2\Big)^{-(\alpha_j+1)/2}, \tag{22}$$

$$Z_j^{-1} = \lambda_j^{-1}\Big(\frac{D_\eta}{D_\xi}\Big)^{\frac{1}{2}}\frac{\Gamma(\frac{1}{2})\Gamma(\frac{\alpha_j}{2})}{\Gamma(\frac{\alpha_j+1}{2})}, \tag{23}$$

where $\alpha_j = -2\nu/(\lambda_j D_\xi) + 1$ (Ito calculus) and $\alpha_j = -2\nu/(\lambda_j D_\xi)$ (Stratonovich calculus). Furthermore the joint probability density function in terms of $x_1, \ldots, x_M$ can be expressed as

$$P(x_1, \ldots, x_M) = \prod_{j=1}^{M}\frac{1}{Z_j}\Big(1 + \lambda_j^2\frac{D_\xi}{D_\eta}\sum_{i=1}^{M}\sum_{k=1}^{M} Q_{ij}Q_{kj}x_i x_k\Big)^{-(\alpha_j+1)/2}. \tag{24}$$

Therefore the M-dimensional q-Gaussian distribution may be expressed as

$$p(x_1,\ldots,x_M)=\prod_{j=1}^{M}\frac{1}{Z_{q_j}(\beta_j)}\Big(1+\beta_j(q_j-1)\sum_{i=1}^{M}\sum_{k=1}^{M}Q_{ij}Q_{kj}x_ix_k\Big)^{\frac{1}{1-q_j}}, \tag{25}$$

by using the unitary matrix Q_{ij} which diagonalizes A_{ij}. Furthermore from Eq. (24) the covariance matrix defined as

$$Cov_{ij}=\int dx_1\ldots\int dx_M x_i x_j P(x_1,\ldots,x_M), \tag{26}$$

is given by

$$Cov_{ij} = \Big(\frac{D_\eta}{D_\xi}\Big)^{\frac{1}{2}}\sum_{k=1}^{M}\lambda_k^{-1}\frac{\Gamma(\frac{1}{2})\Gamma(\frac{\alpha_k}{2})}{\Gamma(\frac{\alpha_k+1}{2})}\int_{-\infty}^{\infty}y_k^2\Big(1+\lambda_k^2\frac{D_\xi}{D_\eta}y_k^2\Big)^{-(\alpha_k+1)/2}dy_kQ_{ik}Q_{kj} \tag{27}$$

$$= \frac{\pi}{2}\Big(\frac{D_\eta}{D_\xi}\Big)^2\sum_{k=1}^{M}\lambda_k^{-2}\frac{\Gamma(\frac{\alpha_k}{2})\Gamma(\frac{\alpha_k}{2}-1)}{\Gamma(\frac{\alpha_k+1}{2})^2}Q_{ik}Q_{kj}. \tag{28}$$

According to the theory on stable distributions [14], the finite variance of stable processes is satisfactory when a characteristic exponent is greater than 2. Hence the expression Eq. (28) is convergent for $\alpha_k > 2 \quad (k=1,\ldots,M)$. Namely, λ_k should be $\lambda_k < 2\nu/D_\xi$ (Ito calculus) and $\lambda_k < \nu/D_\xi$ (Stratonovich calculus). Therefore eigenvalues of covariance matrix Cov_{ij} are given by

$$\mu_k = g(\lambda_k) \equiv \frac{\pi}{2}\Big(\frac{D_\eta}{D_\xi}\Big)^2\lambda_k^{-2}\frac{\Gamma(\frac{\alpha_k}{2})\Gamma(\frac{\alpha_k}{2}-1)}{\Gamma(\frac{\alpha_k+1}{2})^2}\qquad(0<\lambda_k<-a\nu/D_\xi), \tag{29}$$

where $a=2$ (Ito calculus) and $a=1$ (Stratonovich calculus). Eigenvectors corresponding to μ_k are further obtained from the unitary matrix Q_{ij}. Since μ_k is a monotonically increasing function in terms of λ_k, $\mu_k \to 0 \quad (\lambda_k \to 0)$, and $\mu_k \to \infty \quad (\lambda_k \to -a\nu/D_\xi)$ (see. Fig. 1), an inverse function of $g(\lambda_k)$ exists. Here we describe the inverse function of $g(x)$ as $g^{-1}(x)$.

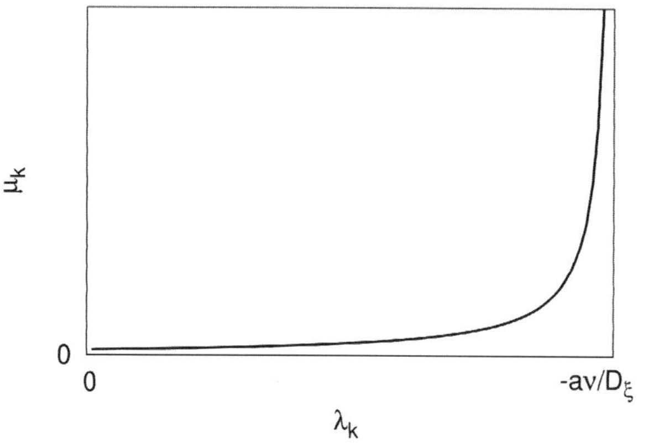

Figure 1. The relationship between μ_k and λ_k.

Practically it is important to estimate A_{ij} from an empirical covariance matrix $\tilde{Cov}_{ij}$ calculated from observed multiple time series with a finite length instead of Cov_{ij}. Let

$r_i(t)$ $(i = 1, \dots, M; t = 1, \dots, T)$ denote multiple financial time series of log returns. Then the empirical covariance matrix is defined as

$$\tilde{Cov}_{ij} = T^{-1} \sum_{t=1}^{T} r_i(t) r_j(t) - \Big(T^{-1} \sum_{t=1}^{T} r_i(t)\Big)\Big(T^{-1} \sum_{t=1}^{T} r_j(t)\Big). \tag{30}$$

One can diagonalize the empirical covariance matrix $\tilde{Cov}_{ij}$ into diagonal matrix $\tilde{\Lambda}_{ij} = \tilde{\mu}_i \delta_{i,j}$ with unitary matrix $\tilde{Q}_{ij}$. Then one has a relation

$$\tilde{\lambda_k} = g^{-1}(\tilde{\mu_k}) \tag{31}$$

and one obtains a relationship between A_{ij} and $\tilde{Cov}_{ij}$

$$A_{ij} = \sum_{k=1}^{M} g^{-1}(\tilde{\mu_k}) \tilde{Q}_{ik} \tilde{Q}_{kj}. \tag{32}$$

4. Portfolio and risks

The portfolio is one of well-known methods to reduce total risks of risky assets by possessing various types of assets. This is mathematically described as follows. The correlation matrix of M risky assets r_i is described as Cov_{ij}, and their means m_i. Then the variance of total assets $R_P(t) = \sum_{i=1}^{M} w_i r_i(t)$ is given by

$$\sigma_P^2 = \sum_{i=1}^{M} \sum_{j=1}^{M} Cov_{ij} w_i w_j, \tag{33}$$

where w_i represents a possessing amount of risky assets i.

Therefore if we know w_i, then we can evaluate the total risk of own portfolio. However this representation has a tendency to evaluate the total risk less than the realized risk due to the Gaussian assumption. Fig. 2 shows q-Gaussian distribution which is calculated from time series generated by Eq. (15) at $M = 1$. The solid red curve represents a normal distribution with the same variance of that q-Gaussian distributions, which is drawn as the blue curve in Fig. 2. It is found that the probability of the q-Gaussian distribution for values greater than 3-σ is higher than one of the normal distribution for values greater than 3-σ. Namely, the variance and the Gaussian assumption can under-evaluate extremal events of risky assets.

Let us consider the probability density function of $R_P(t)$. According to Eq. (32) we can obtain A_{ij} from the empirical time series covariance matrix $\tilde{Cov}_{ij}$. Then we may obtain the distribution of the total assets in portfolio w_i. Since the total assets can be modeled as

$$R_p(t) = \sum_{j=1}^{M} \sum_{i=1}^{M} Q_{ji} w_i y_j(t), \tag{34}$$

they are expressed as a superposition of independent random variables $y_j(t)$ sampled from Eq. (22), i.e. random variables generated from Eq. (19). Hence we can evaluate the default probability of a buffer h with the portfolio w_i from $\mathrm{Prob}\{R_p < -h\}$ by using Eq. (22).

5. Numerical simulations

If one computes empirical covariance matrix $\tilde{Cov}_{ij}$ from multiple financial time series, then one can obtain eigenvalues $\tilde{\mu_k}$ and unitary matrix $\tilde{Q}_{ij}$ by orthgonalizing it. Moreover by using Eq.

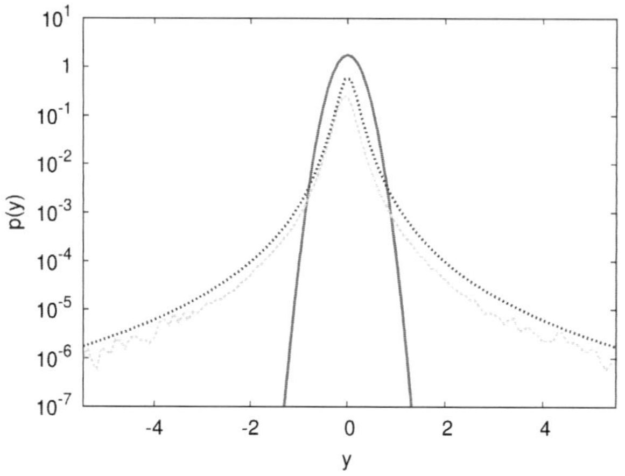

Figure 2. A dashed green curve is the q-Gaussian distribution which is computed from time series obtained from Eq. (15) at $D_\xi = 0.1$, $D_\eta = 2.0$, and $\nu = 2.0$. A dotted blue curve is obtained from Eq. (13) at $D_\xi = 0.1$, $D_\eta = 2.0$, and $\nu = 2.0$. A q-Gaussian distribution and normal distribution. A solid red curve represents zero-mean normal distribution with the same variance as these q-Gaussian distributions.

(32) we can estimate A_{ij} of the multiplicative Langevin equation shown in Eq. (15). Furthermore from Eq. (29) D_η/D_ξ is estimated as

$$\frac{\tilde{D_\xi}}{\tilde{D_\eta}} = \sqrt{\frac{\pi}{2}\frac{1}{\tilde{\lambda_1}^2\tilde{\mu_1}}\frac{\Gamma(\frac{\tilde{\alpha_1}}{2})\Gamma(\frac{\tilde{\alpha_1}}{2}-1)}{\Gamma(\frac{\tilde{\alpha_1}+1}{2})^2}}, \tag{35}$$

where $\tilde{\lambda_1}$ represents the first eigenvalue of the empirical covariance matrix, $\tilde{\mu_1}$ the first eigenvalue of A_{ij}, and $\tilde{\alpha_1}$ the characteristic exponent estimated from the cumulative distribution for the superposition of log returns by the first eigenvector $x_1(t) = \sum_{j=1}^{M} r_j(t)Q_{1j}$. ν/D_ξ is further given by $\nu/D_\xi = -(\alpha_1 - a + 1)/(2\nu)$ at $a = 2$ (Ito calculus) and $a = 1$ (Stratonovich).

Under a given A_{ij} we can generate multiple time series with the same covariance matrix as the empirical estimation and the same power law exponents as of the empirical marginal distribution from Eq. (15). As an example, Fig. 3 shows a sample path which is computed with $A_{ij} = \sum_{l=1}^{M}\sum_{m=1}^{M}\Lambda_{ij}Q_{il}Q_{mj}$ given by a normally distributed random unitary matrix Q_{ij} and $\Lambda_{ij} = \lambda_i\delta_{i,j}$ as uniformly distributed random eigenvalues λ_i ranging from -1 to -0.5 in the case of $M = 2$.

Here we compare a default probability of multiple time series obtained from Eq. (15) under a weight with it obtained from Eq. (34) with the same distributions as their superpositions.

Fig. 4 shows the default probability of the portfolio $w_i = (0.25, 0.5, 0.25)$ in terms of a buffer h from Eq. (15) for $M = 3$. We set A_{ij} as a matrix of which eigenvalues are sampled from a uniform distribution $U(-1, -0.5)$, $\nu = 2.0$, $D_\eta = 0.1$, and $D_\xi = 2.0$. Obviously the default probability is a monotonically decreasing function in terms of the buffer h.

6. Conclusion

We discussed a method to evaluate the total risk of portfolio by means of multi-dimensional q-Gaussian distributions and stochastic differential equations with both multiplicative and additive noises. The multi-dimensional q-Gaussian distributions were derived as a stationary distribution for the stochastic differential equation.

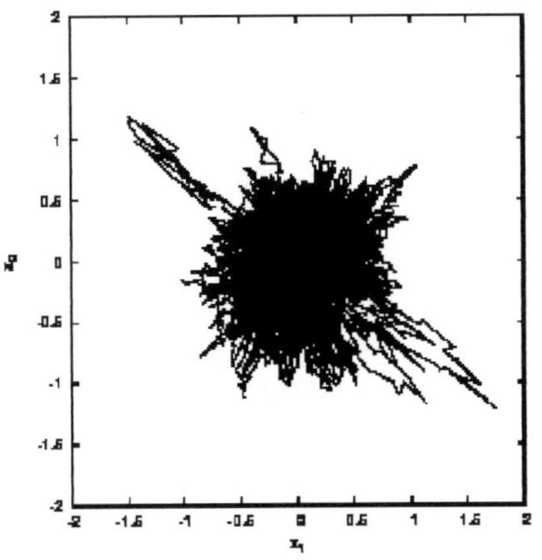

Figure 3. A 2-dimensional scatter plot of dynamical variable obtained from Eq. (15) at $D_\xi = 0.1$, $D_\eta = 2.0$, and $\nu = 2.0$.

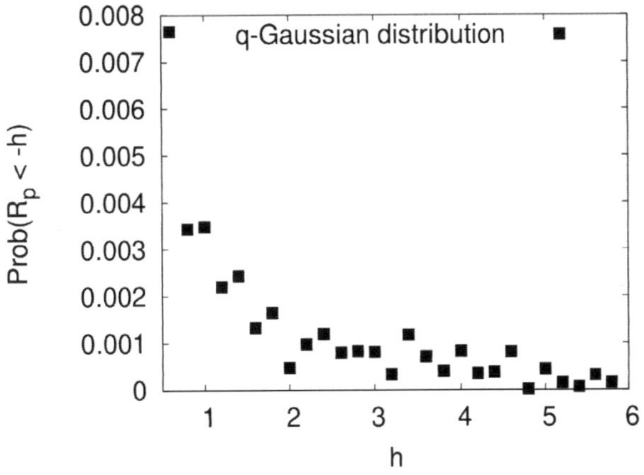

Figure 4. A plot of the default probability in terms of a risk buffer h.

By using the stochastic differential equation with multiplicative and additive noises a method to evaluate the default probability for a given risk buffer under a given portfolio was proposed. From numerical simulations dependence of the default probability on risk buffer size was computed.

As future work empirical computation of default probability should be conducted in order to cope with heavy tailed fluctuations of multiple financial time series by means of the proposed method.

[1] Haken H 1988 *Information and Self-Organization: A Macroscopic Approach to Complex Systems* (Springer-Verlag, Berlin)
[2] Helbing D 2008 *Managing Complexity: Insights, Concepts, Applications* (Springer-Verlag, Berlin)
[3] Sound Credit Risk Assessment and Valuation for Loans - consultative document, Basel Committee on Banking Supervision, issued for comment by 28 February 2006. http://www.bis.org/publ/bcbs121.htm
[4] Mantegna R N and Stanley H E 2000 *An Introduction to Econophysics: Correlations and Complexity in Finance* (Cambridge University Press, Cambridge)

[5] Gell-Mann M and Tsallis C 2004 *Nonextensive Entropy: Interdisciplinary Applications* (Oxford University Press, NY)
[6] Drummond I T 1992 *J. Phys.* A **25** 2273
[7] Nakao H 1998 *Phys. Rev.* E **58** 1591
[8] Kaulakys B and Alaburda M 2006 *J. Stat. Mech.* P02051
[9] Hamilton J D 1989 *Econometrica,* **57** 357
[10] Beck C and Cohen E G D 2003 *Physica* A **322** 110601
[11] Sato A H, Takayasu H, and Sawada Y 2000 *Fractals* **8** 219
[12] Timmermann A 2000 *J. of Econometrics* **96** 75
[13] Kozaki M and Sato A H 2008 *Physica* A **387** 1225
[14] Lévy P 1925 *Calcul des probabités* (Gauthier-Villars, Paris)

Journal of Physics: Conference Series **201** (2010) 012009 doi:10.1088/1742-6596/201/1/012009

Nonequilibrium process of self-gravitating N-body systems and quasi-equilibrium structure using normalized q-expectation values for Tsallis' generalized entropy

Nobuyoshi Komatsu[1], **Shigeo Kimura**[2], **Takahiro Kiwata**[1]

[1]Department of Mechanical Systems Engineering, Kanazawa University, Kakuma-machi, Kanazawa, Ishikawa 920-1192, Japan

[2]The Institute of Nature and Environmental Technology, Kanazawa University, Kakuma-machi, Kanazawa, Ishikawa 920-1192, Japan

E-mail: komatsu@t.kanazawa-u.ac.jp

Abstract. To clarify the nonequilibrium processes of self-gravitating systems, we examine a system enclosed in a spherical container with reflecting walls, by N-body simulations. To simulate nonequilibrium processes, we consider loss of energy through the reflecting wall, i.e., a particle reflected at a non-adiabatic wall is cooled to mimic energy loss. We also consider quasi-equilibrium structures of stellar polytropes to compare with the nonequilibrium process, where the quasi-equilibrium structure is obtained from an extremum-state of Tsallis' entropy. Consequently, we numerically show that, with increasing cooling rates, the dependence of the temperature on energy, i.e., the ε–$\hat{T}$ curve, varies from that of microcanonical ensembles (or isothermal spheres) to a common curve. The common curve appearing in the nonequilibrium process agrees well with an ε–$\hat{T}$ curve for a quasi-equilibrium structure of the stellar polytrope, especially for the polytrope index $n \sim 5$. In fact, for $n > 5$, the stellar polytrope within an adiabatic wall exhibits gravothermal instability [Taruya, Sakagami, Physica A, 322 (2003) 285]. The present study indicates that the stellar polytrope with $n \sim 5$ likely plays an important role in quasi-attractors of the nonequilibrium process in self-gravitating systems with non-adiabatic walls.

1. Introduction

Self-gravitating or long-range attractive interacting systems exhibit several peculiar features, such as negative specific heat and nonequilibrium non-extensive statistical mechanics. For example, negative specific heat plays an important role in astrophysical phenomena, especially in the gravothermal catastrophe of globular clusters [1, 2]. In such systems, standard Boltzmann–Gibbs statistics are not valid because of the existence of long-range potentials. Therefore, non-extensive statistics have been investigated based on Tsallis' generalized entropy [3, 4], by Plastino et al. [5], Abe [6], Chavanis [7] and many other researchers (see e.g. Ref. [4] and references therein).

Recently, from an astrophysical viewpoint, Taruya and Sakagami [8, 9, 10, 11, 12] have elegantly shown that an extremum-state of Tsallis' entropy, i.e., a *stellar polytrope*, has a consistent thermodynamic structure, which implies a thermodynamic instability because of the negative specific heat. They focused on quasi-equilibrium structures of stellar polytropic systems and examined a system enclosed in a spherical container with reflecting walls. This result suggests that the stellar quasi-equilibrium distribution can be described by the stellar polytrope and that the stellar polytropic distribution is expected to be a quasi-attractor of self-gravitating systems. The earlier studies mainly investigated microcanonical ensembles, since a system within an adiabatic wall or the so-called *Antonov problem* [1, 2, 13, 14] is essential for studying self-gravitating systems such as globular clusters.

In fact, from the viewpoint of thermodynamic properties, the influence of energy loss or variation in energy (e.g., the evaporation of stars from globular clusters or radiation loss) has not been examined in detail, with only a few studies [15] considering this influence. Thus it is not yet clear whether the stellar polytrope can be applied to a nonequilibrium process with energy loss and it is therefore important to investigate numerically the thermodynamic properties of the nonequilibrium process with energy loss. In this context, we examine a self-gravitating system enclosed in a spherical container with reflecting walls, by N-body simulations. In this study, a particle reflected at a non-adiabatic wall is cooled to mimic energy loss for nonequilibrium processes. Through this investigation, we attempt to apply the stellar polytrope to the nonequilibrium process with a non-adiabatic wall.

The present paper is organized as follows. In Sec. 2, we give a brief review of numerical techniques for simulating a self-gravitating system enclosed in a spherical container with non-adiabatic walls. In Sec. 3, we briefly review a quasi-equilibrium structure of stellar polytropes calculated from an extremum-state of Tsallis' entropy. In Sec. 4, we present the simulation results and examine the nonequilibrium process with non-adiabatic walls, by a comparison with the stellar polytrope. Finally, we present our conclusions. In this paper, the results shown in Ref. [16] are reconsidered and discussed from the viewpoint of an application of Tsallis' entropy.

2. N-body simulations

We consider a simple system consisting of N point-particles enclosed in a spherical container of radius R with reflecting walls, as shown in Fig. 1. To simulate nonequilibrium processes, we consider the loss of energy through the reflecting wall, i.e., a particle reflected at a non-adiabatic wall is cooled to mimic energy loss. In this section, we briefly review a method to simulate the nonequilibrium process, according to Ref. [16].

To simulate a self-gravitating N-body system, we integrate the set of classical equations of motion for the particles interacting through the Plummer softened potential,

$$\Phi = -\frac{1}{\sqrt{r^2 + r_0^2}}, \tag{1}$$

where r and r_0 represent the distance between particles and the softening parameter, respectively

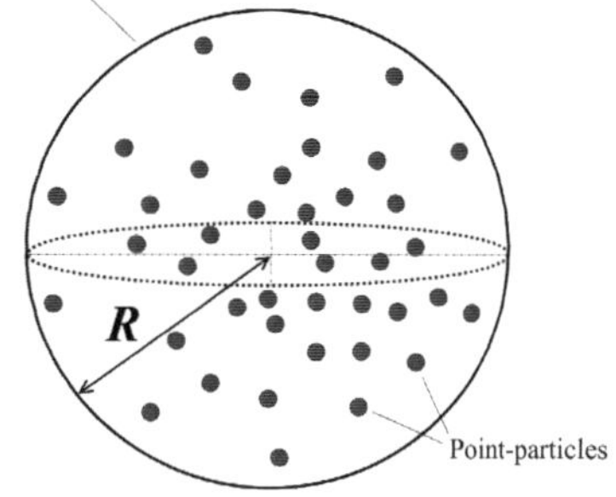

Figure 1. Setup for the N-body simulation. We consider a simple self-gravitating system consisting of N point-particles enclosed in a spherical container of radius R with adiabatic or non-adiabatic walls.

[17, 18, 19]. The total energy E of the self-gravitating N-body system is defined as

$$E = E_{\mathrm{KE}} + E_{\mathrm{PE}} = \sum_{i}^{N} \frac{m_i v_i^2}{2} - \sum_{i<j}^{N} \frac{G m_i m_j}{\sqrt{r_{ij}^2 + r_0^2}}, \tag{2}$$

where E_{KE}, E_{PE} and m_i represent kinetic energy, potential energy and the mass of the i-th point-particle, respectively. G, v_i and r_{ij} represent the gravitational constant, the speed of the i-th particle and the distance between the i-th and j-th particles, respectively. The mass of each particle is set to be m. Applying traditional conventions for self-gravitating systems, the total rescaled energy ε is defined as

$$\varepsilon = \varepsilon_{\mathrm{KE}} + \varepsilon_{\mathrm{PE}} = E\frac{R}{GM^2} = E\frac{R}{G(mN)^2}, \tag{3}$$

where M, $\varepsilon_{\mathrm{KE}}$ and $\varepsilon_{\mathrm{PE}}$ represent the total mass, rescaled kinetic and potential energies, respectively. In this study, the unit of time is $\sqrt{R^3/(Gm)}$. The units are set to be $G = R = m = 1$, to ensure generality of the system. In our units, the temperature $\hat{T}$ of the system is given by

$$\hat{T} = \frac{2}{3k_{\mathrm{B}}}\varepsilon_{\mathrm{KE}} = \frac{2}{3}\varepsilon_{\mathrm{KE}}, \tag{4}$$

assuming that the kinetic energy corresponds to the temperature and that Boltzmann's constant k_{B} is 1.

In this study, we consider a small system consisting of $N = 250$ point-particles in a spherical container of radius $R = 1$. For simulating the N-body system, the set of equations of motion is integrated using Verlet's algorithm. To maintain the accuracy of our simulations, the time step Δt is selected as 10^{-5}, based on a simulation with several different time-steps [17]. Through the present simulations, all interparticle forces are calculated directly at each time step Δt. In this paper, all the numerical results are averaged over 30 simulations with identically prepared initial setups, to observe an averaged behavior of the system.

To mimic a spherical adiabatic wall for microcanonical ensembles, the radial component of the velocity of a particle is reversed when it reaches the wall [17]. Accordingly, the velocity speed $|\boldsymbol{v}_i|$ of the particle is conserved through the reflection. In contrast, to mimic a spherical non-adiabatic wall for nonequilibrium processes, the velocity after the reflection is reduced as

$$\boldsymbol{v}_i^{(\text{reflected})} = (1-\gamma)\boldsymbol{v}_i, \tag{5}$$

where γ is a velocity reduction rate or a cooling rate [16]. That is, a non-adiabatic wall is employed to simulate energy loss by reflecting walls. The cooling rate γ is varied from 0.001 to 0.8, to investigate the influence of the non-adiabatic wall. Note that $\gamma = 0$ corresponds to the adiabatic wall for a microcanonical ensemble. We have checked the total angular momentum through our simulations and confirmed that the variation in total angular momentum is sufficiently small to avoid a phase transition of the system.

For the Plummer softened potential, the softening parameter r_0 is set to be $0.005R$. Accordingly, the collapse and explosion energies for the system with an adiabatic wall [18] are $\varepsilon_{coll} \approx -0.339$ and $\varepsilon_{expl} \approx 0.267$, respectively. In such self-gravitating systems, the crossing time τ_c and the relaxation time τ_r are evaluated as $\tau_c \approx 1/\sqrt{G\rho} = 1/\sqrt{\rho}$ and $\tau_r \approx (0.1N/\ln N)\tau_c$, respectively, where ρ represents the density of the system [13]. In our units, the crossing and relaxation times are $\tau_c \approx 0.1$ and $\tau_r \approx 0.6$, respectively, assuming a uniform density profile. The collapse time of the present system is approximately 600, since the collapse time in a system with $N = 125$–250 particles and $r_0 = 0.005R$ is approximately $\sim 10^3\tau_r$ [19].

For the initial setup, we prepared a self-gravitating system at an approximate virial equilibrium state, by means of 30 microcanonical ensemble simulations in which each simulation time is 10 time units [16]. In the present simulation, the initial total energy ε_0 is set to be 0.5, which is higher than the explosion energy $\varepsilon_{expl} \approx 0.267$. Note that the initial total momentum and initial total angular momentum are set to be 0. To check the system, we examined density profiles and found that the density profile is slightly nonuniform at $\varepsilon_0 = 0.5$, which is generally considered as the uniform state. We also confirmed that the system is in an approximate virial equilibrium state using the virial ratio. Therefore, in our simulations, we employ the above system as the initial setup. The initial energies are $\varepsilon_0 = 0.5000 \pm 0.0096$, $\varepsilon_{\mathrm{KE}_0} = 1.1206 \pm 0.0109$ and $\varepsilon_{\mathrm{PE}_0} = -0.6206 \pm 0.0076$, where the errors indicate the 68% confidence level in terms of the normal error distribution. We have confirmed that our main results do not greatly depend on the initial total energy, e.g., $\varepsilon_0 = 0.3$.

Using the above initial setup, a simulation for a system with a non-adiabatic wall was carried out with various cooling rates γ. For $\gamma = 0.001$, 0.01, 0.1 and 0.2–0.8, the simulation time is $t = 100$, 18, 4 and 2, respectively. Accordingly, our simulation time should be shorter than the collapse time of the present system, ~ 600; that is, we examine an early relaxation process before the collapse. As for $\gamma = 0$, a microcanonical ensemble simulation is carried out.

3. Stellar polytrope

Taruya and Sakagami [8, 9, 10, 11, 12] have shown that the extremum-state of Tsallis' entropy, i.e., the *stellar polytrope*, has a consistent thermodynamic structure, which implies thermodynamic instability due to a negative specific heat. They found that the stellar polytropic distribution should be a quasi-attractor of self-gravitating systems. We expect that the quasi-equilibrium structure of the stellar polytrope should be related to a nonequilibrium process appearing in the self-gravitating system with a non-adiabatic wall. Therefore, we briefly review the stellar polytrope calculated from an extremum-state of Tsallis' entropy, according to the works of Taruya and Sakagami [11, 12].

3.1. Tsallis' entropy and stellar polytrope

In the new framework of Tsallis' non-extensive thermo statistics, all the macroscopic observables of quasi-equilibrium systems should be characterized by escort distributions, although the escort distributions themselves should not be fundamental [11, 12]. In fact, we can find a more fundamental probability function $p(\boldsymbol{x}, \boldsymbol{v})$ that quantifies the phase-space structure, where $(\boldsymbol{x}, \boldsymbol{v})$ represents 6-dimensional phase-space. Note that the phase-space measure is defined as $\mathrm{d}^6\boldsymbol{\tau} \equiv \mathrm{d}^3\boldsymbol{x}\mathrm{d}^3\boldsymbol{v}/h^3$, where $h^3 \equiv (l_0 v_0)^3$ represents the phase-space element with the unit of length l_0 and the unit of velocity v_0.

Using the fundamental probability function $p(\boldsymbol{x}, \boldsymbol{v})$, the escort distribution is defined as

$$P_q(\boldsymbol{x}, \boldsymbol{v}) = \frac{\{p(\boldsymbol{x}, \boldsymbol{v})\}^q}{\int \mathrm{d}^6\boldsymbol{\tau}\{p(\boldsymbol{x}, \boldsymbol{v})\}^q}, \tag{6}$$

and the macroscopic observables expressed as normalized q-expectation values are given by

$$\langle O_i \rangle_q = \int \mathrm{d}^6\boldsymbol{\tau}\{O_i P_q(\boldsymbol{x}, \boldsymbol{v})\}, \tag{7}$$

where the normalization condition for $p(\boldsymbol{x}, \boldsymbol{v})$ is given as

$$\int \mathrm{d}^6\boldsymbol{\tau}\{p(\boldsymbol{x}, \boldsymbol{v})\} = 1. \tag{8}$$

Tsallis' generalized entropy is given as

$$S_q = -\frac{1}{q-1}\int \mathrm{d}^6\boldsymbol{\tau}[\{p(\boldsymbol{x},\boldsymbol{v})\}^q - p(\boldsymbol{x},\boldsymbol{v})]. \tag{9}$$

Through an extremum-state of Tsallis' entropy, i.e., solving the variational problem, Taruya and Sakagami have shown that the polytropic relation can be given as

$$P(r) = K_n\{\rho(r)\}^{1+1/n}, \tag{10}$$

where K_n, $P(r)$ and $\rho(r)$ are the dimensional constant, the isotropic pressure and the density at radius r, respectively [11, 12]. The polytrope index n is given by

$$n = \frac{1}{1-q} + \frac{1}{2}. \tag{11}$$

3.2. *Emden solutions and thermodynamic properties of the stellar polytrope*

In this section, we focus on a spherically symmetric system with polytrope index $n > 3/2$, in which the equilibrium distribution is at least *dynamically* stable [13]. In such situations, the stellar equilibrium distribution should be characterized by the so-called *Emden solutions*, and all the physical quantities should be discussed using the homology invariant variables [12]. For example, Chavanis examined the gravitational instability of isothermal and polytropic spheres in detail, using the Milne variables [7, 22, 23]. Note that we consider here a quasi-equilibrium structure of the stellar polytropic system with pure gravitational potentials [11, 12].

The equations for the hydrostatic equilibrium of self-gravitating systems [11, 12] are given by

$$\frac{\mathrm{d}P(r)}{\mathrm{d}r} = -\frac{Gm(r)}{r^2}\rho(r), \tag{12}$$

$$\frac{\mathrm{d}m(r)}{\mathrm{d}r} = 4\pi\rho(r)r^2, \tag{13}$$

where $m(r)$ represents the mass evaluated at radius r. The dimensionless quantities, i.e., θ and ξ, are introduced as

$$\rho = \rho_c[\theta(\xi)]^n, r = \left\{\frac{(n+1)P_c}{4\pi G{\rho_c}^2}\right\}^{1/2}\xi, \tag{14}$$

where ρ_c and P_c are the density and pressure at the center of the spherical container, respectively. Then, the ordinary differential equation for the hydrostatic equilibrium is given as

$$\theta'' + \frac{2}{\xi}\theta' + \theta^n = 0, \tag{15}$$

and the boundary conditions are set to be

$$\theta(0) = 1, \theta'(0) = 0, \tag{16}$$

where the prime represents the derivative with respect to ξ, i.e., $\mathrm{d}/\mathrm{d}\xi$. Solutions of Eq. (15) with (16), i.e., Emden solutions, are calculated numerically using, for example, Runge–Kutta methods. This is because the solution cannot be solved analytically, except for the polytrope indices, $n = 0$, 1 and 5.

Journal of Physics: Conference Series **201** (2010) 012009 doi:10.1088/1742-6596/201/1/012009

To characterize the equilibrium properties of the Emden solution [11, 12], homology invariants are defined as

$$u \equiv \frac{\mathrm{d}\ln m(r)}{\mathrm{d}\ln r} = \frac{4\pi r^3 \rho(r)}{m(r)} = -\frac{\xi\theta^n}{\theta'}, \tag{17}$$

$$v \equiv -\frac{\mathrm{d}\ln P(r)}{\mathrm{d}\ln r} = \frac{\rho(r)}{P(r)}\frac{Gm(r)}{r} = -(n+1)\frac{\xi\theta'}{\theta}. \tag{18}$$

In the following, we consider a quasi-equilibrium structure of the stellar polytropic system enclosed in a spherical container with adiabatic walls. Taruya and Sakagami have evaluated the total energy E of a confined stellar system in terms of the total mass M, the pressure P_e and density ρ_e at the boundary r_e. (They considered an extremum-state of Tsallis' entropy at a fixed mass and energy.) Moreover, they determined a plausible physical temperature T_{phys}, using the modified Clausius relation [11, 12]. Based on this analysis, as a function of the homology invariants at the wall, i.e., u_e and v_e, the dimensionless quantities for the total energy and inverse temperature are given by

$$\lambda \equiv -\frac{Er_e}{GM^2} = -\frac{1}{n-5}\left\{\frac{3}{2}\left(1-\frac{n+1}{v_e}\right) + (n-2)\frac{u_e}{v_e}\right\}, \tag{19}$$

$$\eta \equiv \frac{GM^2}{r_e T_{phys}} = \frac{(n-5)v_e}{n+1-2u_e-v_e}, \tag{20}$$

where r_e corresponds to R in the present study. For details, see Eqs. (37) and (42) in Ref. [11] or Eqs. (3.27) and (3.33) in Ref. [12].

In our units, the total rescaled energy and temperature can be given as

$$\varepsilon = -\lambda, \tag{21}$$

$$\hat{T} = 1/\eta. \tag{22}$$

To confirm the thermodynamic properties of the stellar polytrope, we observe the trajectories of the Emden solutions in the ε–$\hat{T}$ plane. As shown in Fig. 2, for higher energies, the temperature $\hat{T}$ decreases with decreasing total energy ε. Accordingly, the system behaves like an ideal gas with a positive specific heat, i.e., $d\varepsilon/d\hat{T} > 0$. In contrast, for lower energies, the temperature increases as the total energy decreases. In other words, the specific heat in this region is negative, i.e., $d\varepsilon/d\hat{T} < 0$. This is because the potential energy of the system is dominant in the lower energy

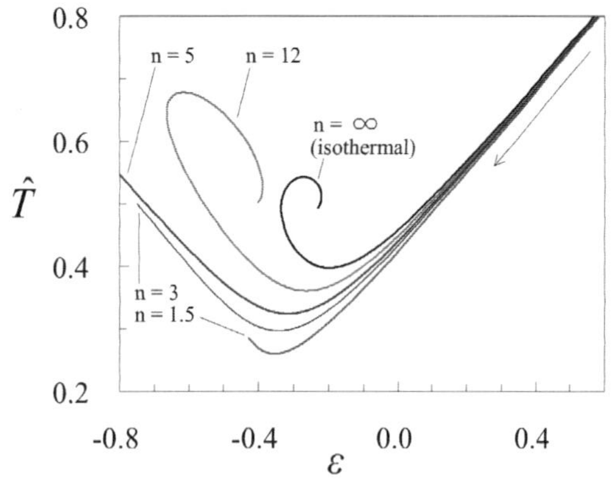

Figure 2. (Color online) Trajectories of Emden solutions in the ε–$\hat{T}$ plane. Each point along the trajectory represents an Emden solution evaluated at different radiuses r_e. Based on the boundary conditions, all the trajectories start from $(\varepsilon,\ \hat{T}) = (\infty,\ \infty)$ corresponding to the limit $r_e \to 0$. With increasing radius, the trajectories move as indicated by the arrow. The polytrope index $n = \infty$ corresponds to an isothermal sphere [16].

region. We now observe the influence of the polytrope index n. We find that the trajectories for $n > 5$ gradually change their direction and finally spiral around a fixed point, while the trajectories for $n < 5$ terminate suddenly because of the finite radius [11]. In fact, for $n > 5$, a stellar polytrope within an adiabatic wall exhibits gravothermal instability, as investigated by Taruya and Sakagami [11, 12].

4. Results

We first examine the nonequilibrium process with non-adiabatic walls by N-body simulations. For this, we observe typical time evolutions of the energies for $\gamma = 0.001$, 0.01, 0.1, 0.2, 0.3, 0.4 and 0.5. As shown in Fig. 3, the total energy ε decreases rapidly from the initial total energy of $\varepsilon_0 = 0.5$, due to the non-adiabatic wall. Hereafter we call the first stage the *rapid initial cooling stage* [20]. We can confirm that the behavior of each curve is consistent, especially during the rapid initial cooling stage. Similarly, the kinetic energy $\varepsilon_{\mathrm{KE}}$ decreases with γt during the rapid initial cooling stage. However, the kinetic energy gradually increases after this stage, while the total energy decreases, i.e., $d\varepsilon/d\varepsilon_{\mathrm{KE}}$ is negative. This indicates that an incidence of negative specific heat occurs, assuming that the kinetic energy corresponds to the temperature of the system [16].

For $\gamma = 0.3$, 0.4 and 0.5, the kinetic energy oscillates considerably during the rapid initial cooling stage, after which the amplitude of the oscillations decays. This is because, for larger cooling rates, the decrease in energy is too large to maintain the virial equilibrium state during the rapid initial cooling stage [16]. However, all the systems are in an approximate virial equilibrium state, except around the large oscillations and the rapid initial cooling stage. We have confirmed that the oscillation does not greatly depend on the average operation and the time step. The oscillation may disappear when the number of particles is large [21].

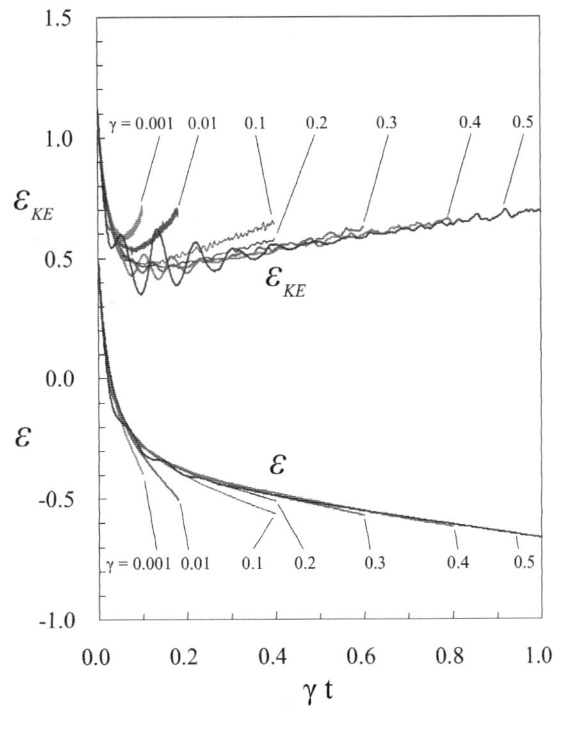

Figure 3. (Color online) Time evolutions of the total energy ε and kinetic energy $\varepsilon_{\mathrm{KE}}$, for various cooling rates γ. The horizontal axis represents the rescaled time γt. The results are averaged over 30 simulations with identically prepared initial setups [16].

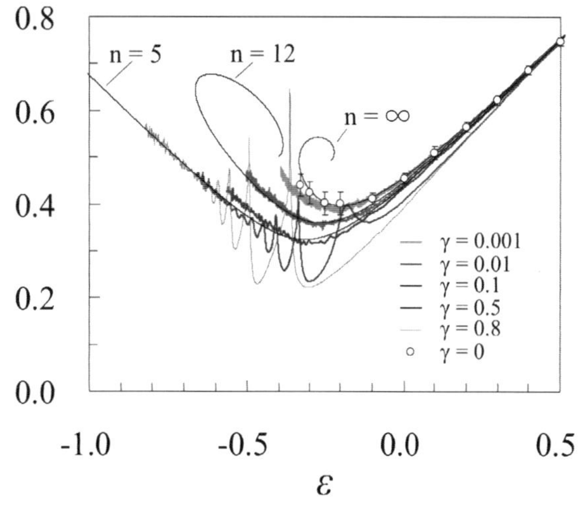

Figure 4. (Color online) Dependence of the temperature $\hat{T}$ on the total energy ε for various cooling rates γ and polytrope indices n. For $\gamma > 0$, the simulation starts from an initial total energy of $\varepsilon_0 = 0.5$. The open circles represent $\gamma = 0$, i.e., microcanonical ensemble simulations [16]. The trajectories of Emden solutions with $n = 5$, 12 and ∞ are re-plotted from Fig. 2.

Next, to examine the nonequilibrium processes of the self-gravitating system, we observe the dependence of the temperature $\hat{T}$ on the total energy ε for various cooling rates, as shown in Fig. 4. In this figure, we calculate the temperature $\hat{T}$ from the kinetic energy using Eq. (4). (The assumption that the kinetic energy corresponds to the temperature of the system is valid, since all the systems are in an approximate virial equilibrium state, except around the large oscillations and the rapid initial cooling stage [16].) Moreover, in Fig. 4, we re-plot the trajectories of the Emden solutions for the polytrope indices $n = 5$, 12 and ∞, to compare with the nonequilibrium process for various cooling rates. Note that a pure gravitational potential is used for the stellar polytrope, while the softening parameter is set to be $r_0 = 0.005R$ in our nonequilibrium simulations. Accordingly, we have to take into account the influence of the softening parameter, since it affects the dependence of the temperature on the total energy.

As shown in Fig. 4, our microcanonical ensemble simulations ($\gamma = 0$) are consistent with the isothermal sphere ($n = \infty$), except for a part of the spiral curve. The difference between $\gamma = 0$ and $n = \infty$ indicates the influence of the deviation from a pure gravitational potential. For $-0.2 \lessapprox \varepsilon$, the system behaves like an ideal gas with a positive specific heat, while the specific heat for $\varepsilon \lessapprox -0.2$ is negative [16]. To compare with the microcanonical ensemble, we examine a system with a small cooling rate, $\gamma = 0.001$. (The simulations for $\gamma > 0$ start from an initial total energy of $\varepsilon_0 = 0.5$.) As expected, the result for $\gamma = 0.001$ agrees well with the microcanonical ensemble ($\gamma = 0$), although the temperature for $\gamma = 0.001$ is slightly lower than that for $\gamma = 0$. We can confirm that a negative specific heat occurs not only in the microcanonical ensemble but also in certain nonequilibrium processes with a non-adiabatic wall.

Now, we investigate the influence of the cooling rate more closely. As shown in Fig. 4, for $-0.2 \lessapprox \varepsilon$, the curve further deviates from the microcanonical ensemble ($\gamma = 0$) with increasing cooling rate γ. In other words, the greater the cooling rate, the steeper the slope of the curve. For $\varepsilon \lessapprox -0.2$, or rather $\varepsilon \lessapprox \varepsilon_{coll}$, the temperature increases with decreasing total energy, excepting several large oscillations. (All the systems are in an approximate virial equilibrium state, except around the large oscillations and the rapid initial cooling stage.) Therefore, it is clearly demonstrated that a negative specific heat occurs in the system for various cooling rates. Moreover, each curve gradually shifts towards a certain common curve. However, the extent of the negative specific heat or $d\varepsilon/d\hat{T}$ does not greatly depend on the cooling rate. That is, with increasing cooling rate, the dependence of the temperature on energy, i.e., the ε–$\hat{T}$ curve, varies from the microcanonical ensemble to a common curve.

We expect that the tendency towards a common curve is related to a kind of quasi-attractor

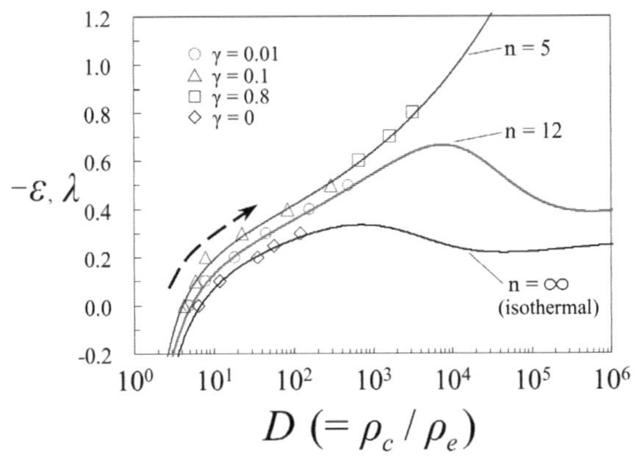

Figure 5. (Color online) Density contrast $D(=\rho_c/\rho_e)$ and the negative of the energy $\lambda(=-\varepsilon)$. The real lines represent the stellar polytropes with $n=5$, 12 and ∞. The symbols represent the simulation results of the nonequilibrium process for various cooling rates γ. For $\gamma>0$, the symbols represent evolutionary states; that is, the states for $\gamma>0$ evolve as indicated by the arrow, as time progresses.

discussed in Ref. [8] by Taruya and Sakagami. (Note that the stellar polytrope examined by Taruya and Sakagami is assumed to be a quasi-equilibrium structure under the restriction of constant total energy, while the total energy is not fixed in the present nonequilibrium simulation.) Accordingly, we compare our nonequilibrium process with a quasi-equilibrium structure of stellar polytropes. We find that the common curve appearing in the nonequilibrium process agrees well with an ε–$\hat{T}$ curve for the polytrope index of $n=5$. (The polytrope index of $n=5$ corresponds to the appearance of gravothermal instability for a stellar polytrope within an adiabatic wall [11, 12].) This indicates that a quasi-equilibrium structure for a stellar polytrope with $n\sim5$ should play an important role in quasi-attractors of the present nonequilibrium simulation. (We have confirmed that the tendency towards a common curve does not greatly depend on the softening parameter, e.g., $r_0=0.050R$. Of course, the deviation from the ε–$\hat{T}$ curve for $n=5$ depends on the softening parameter.) Moreover, we found that the curve for $\gamma=0.01$ agrees well with the curve for a polytrope index of $n=12$, except for a part of the spiral curve.

As discussed above, the present nonequilibrium process is expected to be related to the stellar polytrope. To examine this more closely, we observe the relationship between the density contrast $D=\rho_c/\rho_e$ and the negative of the total energy $\lambda(=-\varepsilon)$, for various cooling rates γ and polytrope indices n. (Here ρ_c and ρ_e represent the densities at the center and wall, respectively. Based on the works of Taruya and Sakagami [11, 12], $\lambda(=-\varepsilon)$ is used in this discussion.) To obtain the density contrast for various cooling rates, the density profile calculated from the Emden solution is fitted with the corresponding density profile of our simulations, using Eq. (14). Based on the fitted curves, we evaluated the density contrast for various cooling rates. In Fig. 5, typical density contrasts are plotted except for large oscillations of the large cooling rates. Note that, for $\gamma>0$, the symbols in the figure represent evolutionary states; that is, the states for $\gamma>0$ evolve as indicated by the arrow, as time progresses.

As shown in Fig. 5, the results for $\gamma=0$ (◇) and $\gamma=0.01$ (○) agree well with the D–λ curves for $n=\infty$ and $n=12$, respectively. As expected, this is consistent with the result shown in Fig. 4. Similarly, the results for $\gamma=0.1$ (△) and $\gamma=0.8$ (□) agree well with the D–λ curve for $n=5$. That is, the common curve appearing in $\gamma=0.1$–0.8 is likely to be related to stellar polytrope with $n\sim5$. Of course, the stellar polytrope is assumed to be a quasi-equilibrium structure under the restriction of constant total energy. However, these results indicate that the stellar polytrope calculated from Tsallis' entropy can be applied to the present nonequilibrium process with a non-adiabatic wall.

5. Conclusions

To clarify the nonequilibrium processes of self-gravitating systems, we examined a system enclosed in a spherical container with reflecting walls, by N-body simulations. For this, we investigated the nonequilibrium process numerically and compared it with a quasi-equilibrium structure of stellar polytropes, which is calculated from an extremum-state of Tsallis' generalized entropy. In our simulations, for microcanonical ensembles, the radial component of the velocity of a particle was reversed when it reached the wall. In contrast, for simulating a nonequilibrium process, we considered loss of energy through the reflecting wall, i.e., a particle reflected at a non-adiabatic wall was cooled to mimic energy loss. We have shown that, with increasing cooling rates, the dependence of the temperature on energy, i.e., the ε–$\hat{T}$ curve, varies from the microcanonical ensemble to a common curve. The common curve appearing in the nonequilibrium process is consistent with an ε–$\hat{T}$ curve for the polytrope index $n = 5$ corresponding to the appearance of gravothermal instability. This indicates that the quasi-equilibrium structure of stellar polytropes for $n \sim 5$ should play an important role in quasi-attractors of the nonequilibrium process appearing in a self-gravitating system with a non-adiabatic wall. The stellar polytrope considered here is assumed to be a quasi-equilibrium structure under the restriction of constant total energy. However, we found that the stellar polytrope calculated from Tsallis' entropy can be applied to the nonequilibrium process with a non-adiabatic wall. That is, the behavior of the present nonequilibrium process is likely to be related to the stellar polytrope.

We are grateful to Prof. M. Sakagami for valuable comments in the Kyoto RIMS workshop.

References

[1] Antonov V A 1962 Vestn. Leningrad Univ. **7** 135; *in Dynamics of Globular Clusters*, edited by Goodman J and Hut P 1985 *IAU Symposium* No. 113 (Reidel, Dordrecht)
[2] Lynden-Bell D and Wood R 1968 *Mon. Not. R. Astron. Soc.* **138** 495
[3] Tsallis C 1988 *J. Stat. Phys.* **52** 479
[4] Tsallis C 2009 *Introduction to Nonextensive Statistical Mechanics: Approaching a Complex World* (Springer)
[5] Plastino A and Plastino A R 1993 *Phys. Lett. A* **174** 384
[6] Abe S 1999 *Phys. Lett. A* **263** 424
[7] Chavanis P H 2002 *Astron. Astrophys.* **386** 732
[8] Taruya A and Sakagami M 2003 *Phys. Rev. Lett.* **90** 181101
[9] Taruya A and Sakagami M 2005 *Mon. Not. R. Astron. Soc.* **364** 990
[10] Taruya A and Sakagami M 2002 *Physica (Amsterdam)* **307A** 185; 2003 *Physica (Amsterdam)* **318A** 387
[11] Taruya A and Sakagami M 2003 *Physica (Amsterdam)* **322A** 285
[12] Sakagami M and Taruya A 2004 *Continuum Mech. Thermodyn.* **16** 279
[13] Binney J and Tremaine S 1987 *Galactic Dynamics* (Princeton University Press, Princeton)
[14] Padmanabhan T 1990 *Phys. Rep.* **188** 285
[15] Posch H A and Thirring W 2005 *Phys. Rev. Lett.* **95** 251101
[16] Komatsu N, Kimura S and Kiwata T 2009 *Phys. Rev. E* (in print)
[17] Komatsu N, Kiwata T and Kimura S 2008 *Physica (Amsterdam)* **387A** 2267; 2009 *Physica (Amsterdam)* **388A** 639
[18] Ispolatov I and Karttunen M 2003 *Phys. Rev. E* **68** 036117
[19] Ispolatov I and Karttunen M 2004 *Phys. Rev. E* **70** 026102
[20] Komatsu N and Abe T 2005 *Phys. Rev. E* **72** 021601
[21] Merrall T E C and Henriksen R N 2003 *Astrophys. J.* **595** 43
[22] Chavanis P H 2002 *Astron. Astrophys.* **381** 340
[23] Chavanis P H 2003 *Astron. Astrophys.* **401** 15

Journal of Physics: Conference Series **201** (2010) 012010 doi:10.1088/1742-6596/201/1/012010

Experimental investigation on the surface potential decays of dielectric materials with q-exponential function

Junsei Horikawa

Graduate School of Science and Engineering, Ibaraki University, Ibaraki, 316-8511, Japan

E-mail: 09nm621n@hcs.ibaraki.ac.jp

Tatsuaki Wada

Department of Electrical and Electronic Engineering, Ibaraki University, Ibaraki, 316-8511, Japan

E-mail: wada@mx.ibaraki.ac.jp

Abstract. We have studied the surface potential decays (SPD) of a variety of dielectric materials, and found that the SPD data of the samples with high charge retensions are well fitted by Tsallis q-exponential functions.

1. Introduction

In many different fields we have seen Tallis q-exponential distributions, which is equivalent to the Zipf-Mandelbrot distributions in the case with the parameter q is lager than unity. For example, the distributions of family names [1], urban agglomeration [2], itineration of the internet over nonequilibrium stationary states [3], fluxes of cosmic ray [4], and football goals [5].

In this article, we show that the surface potential decay (SPD) profiles of various dielectric material powders are well fitted by Tsallis q-exponential functions. SPD measurement is a conventional experimental technique [6] for studying charge carrier transport and charge retention in dielectric (or insulating) materials. The SPD technique has been widely applied many kinds of dielectric materials concerning the industrial fields such as electro-photography or electrostatic powder coating or insulators for electric cable. A characteristic property of the SPD curves of dielectric materials with high electrical charge retention is an exponential behavior in the early stage and power law behavior in final stage as will be shown in figure 3 in section 2. Because of this property, the SPD profiles of dielectric materials with high electrical charge retention have not been well fitted by a single analytical function yet. To the best of our knowledge, no such reports are available.

The organization of the paper is as follows. In section 2, we explain the samples and the experimental procedure for SPD measurements. Then Tsallis q-exponential function and fitting procedure are explained. Section 3 shows the results and discussion. Remarkably the most of the SPD profiles are well fitted by Tsallis q-exponential functions. Finally, the paper is concluded in section 4.

Journal of Physics: Conference Series **201** (2010) 012010 doi:10.1088/1742-6596/201/1/012010

2. Experimental and fitting procedures

2.1. Samples

Table 1 shows the dielectric materials used in this experiment, and their mean diameters. Polyester is used as a raw material for a PET bottle and fiber clothes. Epoxy is used by engineer plastic. Acrylic is used as a substitute of glass. Nylon is a kind of polyamide, and used as raw materials for clothes. Polyvinyl chloride powder, polyester powder, epoxy powder, and acrylic powder are used for electrostatic powder coating. Ferrite carriers is used as carrier particles in electro-photography (or Xerography). Those samples are high electrical charge retention. Nylon powder is used for powder coating.

Table 1. Sample materials

Sample name	Mean diameter (μm)	Uses
Ferrite carriers	55	electro-photography
Polyvinyl chloride powder	200	electrostatic powder coating
Polyester powder	27	electrostatic powder coating
Epoxy powder	20	electrostatic powder coating
Acrylic powder	10	electrostatic powder coating
Nylon powder	15	powder coating

2.2. Experimental procedure

Figure 1 show our experimental set up for DC corona-charging of samples. A sample is filled in the Aluminium container, which is 3 mm in depth and 30 mm in the diameter as shown figure 1. The sample container and grid mesh are electrically grounded. The sample is charged by DC corona discharging. For positive charging +5kV is applied to the tip electrode, and the grid potential is set to about +500 V. For negative charging the polarity of the applied voltages are reversed. The distance between the grid mesh and the surface of the sample is set to 5 mm. The sample is exposed to the DC corona discharge for about 1 min.

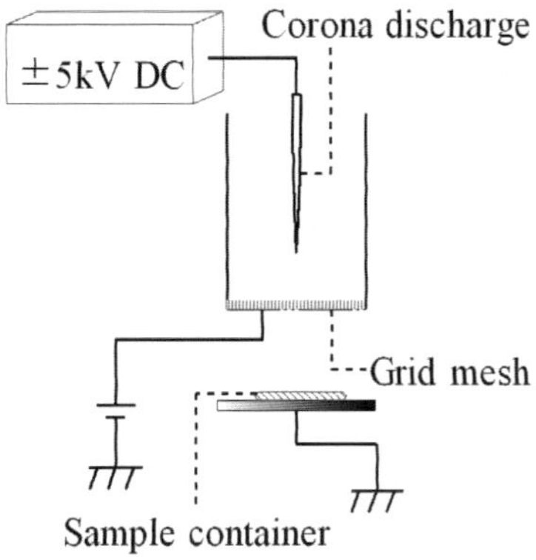

Figure 1. Corona electrification experimental setup

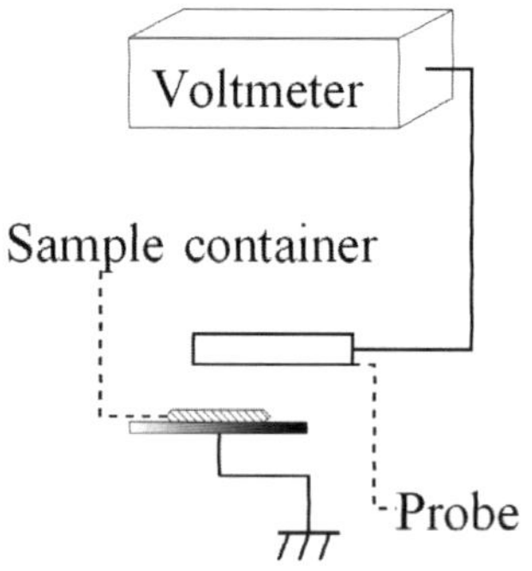

Figure 2. Measurement experimental setup

After the corona charging the sample is immediately set to the under the measurement probe as shown in figure 2, and the surface potential is measured with an electro static voltmeter (TREK-344). The measurement probe (TREK MODEL 600B-8) is located in 3 mm from the sample surface.

2.3. Fitting procedure

A characteristic property of the SPD curves of dielectric materials with high electrical charge retention is an exponential behavior in the early stage and power law behavior in final stage as shown in figure 3, which is a typical SPD profile of ferrite carriers. Because of this property,

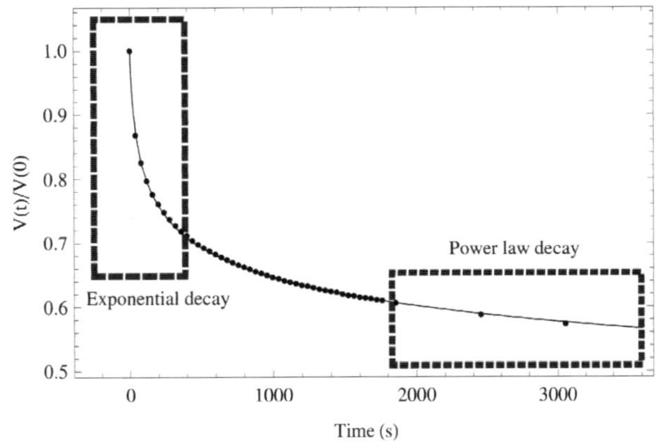

Figure 3. A typical SPD profile of ferrite carriers.

the SPD profiles of dielectric materials with high electrical charge retention have not been well fitted by a single analytical function yet. An analytic function which has such property is Tsallis q-exponential function [7, 8], which is defined by

$$\exp_q(x) \equiv [1 + (1-q)x]^{\frac{1}{1-q}}, \tag{1}$$

for a real parameter $q > 1$. In the limit of $q \rightarrow 1$, the q-exponential function reduces to $\exp(x)$. Since this q-exponential function behaves as

$$\exp_q(x) \sim \exp(x), \tag{2}$$

for a small value of $|x|$, and

$$\exp_q(x) \sim |x|^{-\frac{1}{q-1}} \tag{3}$$

for a large value of $|x|$, we use this analytic function in order to fit the SPD data of the samples, especially with high charge retention. The measured SPD data is therefore fitted by using the following function

$$U(t) = U(0)\exp_q\left(-\frac{t}{\tau}\right), \tag{4}$$

where q and τ are fitting parameters, $U(t)$ is a surface potential at time t, $U(0)$ is an initial surface potential just starting measurement $t = 0$.

In order to judge whether the fitting is well or not, we use the sum of the squares of the residuals (divided by the number) R^2 defined by

$$R^2 = \frac{1}{N}\sum_{i=1}^{N}\{V(t_i) - U(t_i)\}^2 . \tag{5}$$

Here N is the number of the measured data, $V(t_i)$ and $U(t_i)$ are a measured data and the value of the fitting function at $t = t_i$, respectively. The smaller R^2 the better the fitting.
The q-logarithmic function is defined by

$$\ln_q(x) = \frac{x^{1-q} - 1}{1-q}, \tag{6}$$

which is the inverse function of $\exp_q(x)$. By taking q-logarithmic function of the both sides of equation (4), we have

$$\ln_q\left[\frac{U(t)}{U(0)}\right] = -\frac{t}{\tau}. \tag{7}$$

Then, the q-logarithmic functions of the normalized SPD data obey a straight line if the data is well fitted by the q-exponential function. Such a plot is called the q-logarithmic plot, which is as useful as the standard logarithmic plot.

3. Results and discussion

The powder samples for electrostatic coating and ferrite carrier have high charge retention characteristics, and consequently the SPD of those samples are relatively slow. Whereas nylon powder have low charge retention characteristics, and consequently the SPD of those samples are relatively fast.

3.1. Powders for electrostatic powder coating

Figures 4-7 are results of the SPD profiles of polyvinyl chloride powder, acrylic powder, epoxy powder, and polyester powder, respectively Each SPD curve is plotted with the surface potential normalized by initial surface potential $V(0)$ against the total measurement time.

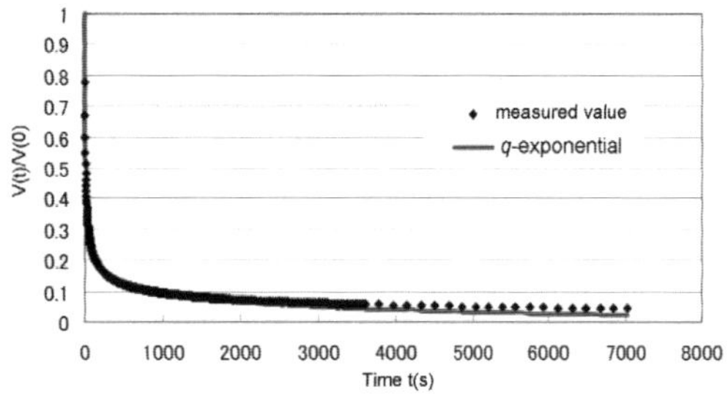

Figure 4. SPD profile of polyvinyl chloride powder. Relative humidity is 47% and temperature is 25.9°C. The total measurement time is 7000s. Parameter q is 3.59, parameter τ is 5.53 and R^2 is 2.85×10^{-5}

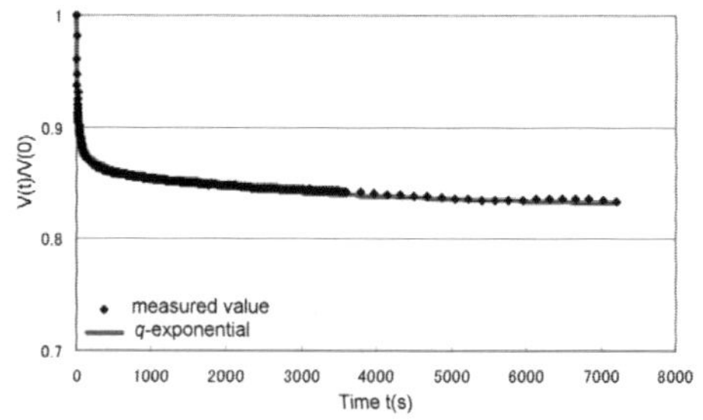

Figure 5. SPD profile of acrylic powder. Relative humidity is 60% and 25.5°C. The total measurement time is 7000s. Parameter q is 76.2, parameter τ is 0.52 and R^2 is 1.34×10^{-5}

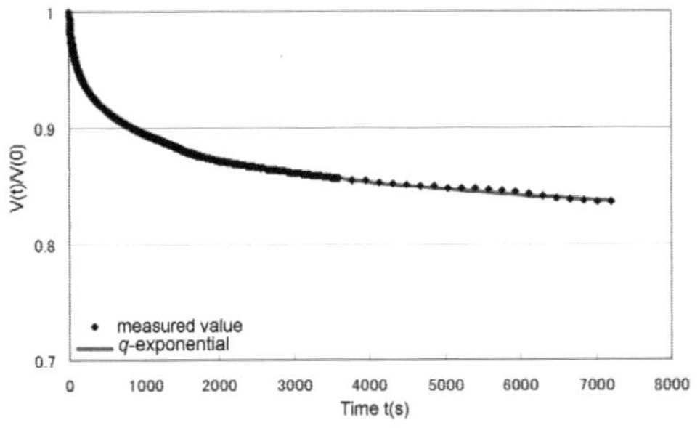

Figure 6. SPD profile of epoxy powder. Relative humidity is 61% and temperature is 25.4°C. The total measurement time is 7000s. Parameter q is 29.7, parameter τ is 16.5×10^2 and R^2 is 1.29×10^{-6}

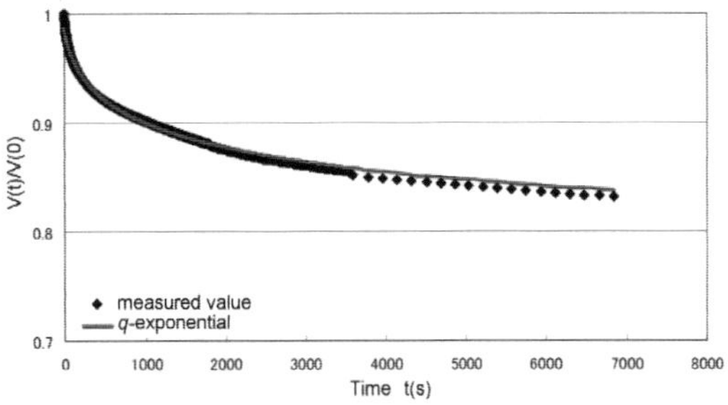

Figure 7. SPD profile of polyester powder. Relative humidity is 56% and temperature is 24.5°C. The total measurement time is 7000s. Parameter q is 26.8, parameter τ is $26.0 \times 10^{-6}$and R^2 is 4.09×10^{-6}

The normalized surface potential $V(t)/V(0)$ soon decays to a certain value in the early stage, and then slow decay later for the result of SPD profile of polyvinyl chloride powder and that of acrylic powder. These properties are reflected in a small value of the fitting parameter τ, which is a similar quantity as a time constant in the standard exponential decays. However τ in the q-exponential decay depends on q. As obvious from equation (7), at $t = \tau$ the value of the normalized surface potential decreases by the factor of

$$\exp_q(-1) = q^{\frac{1}{1-q}}. \qquad (8)$$

3.2. Ferrite carriers for electro-photography

Figure 8 is a result of the SPD profile (dot) of ferrite carriers. Initial surface potential $V(0)$ is -496.8 V and the total measurement time is 6000 s.

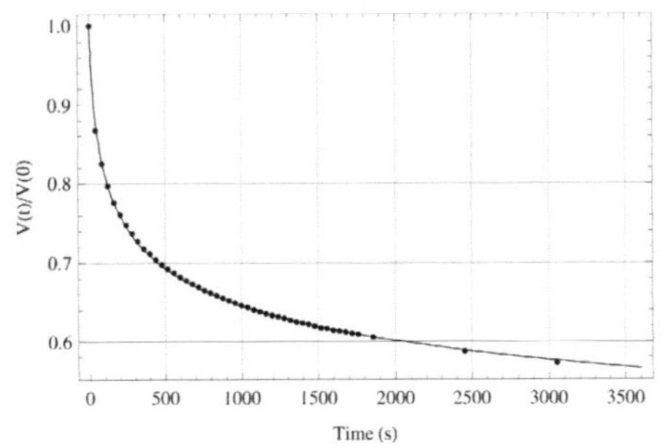

Figure 8. SPD profile of ferrite carrier. Relative humidity is 57% and temperature is 27.8°C.

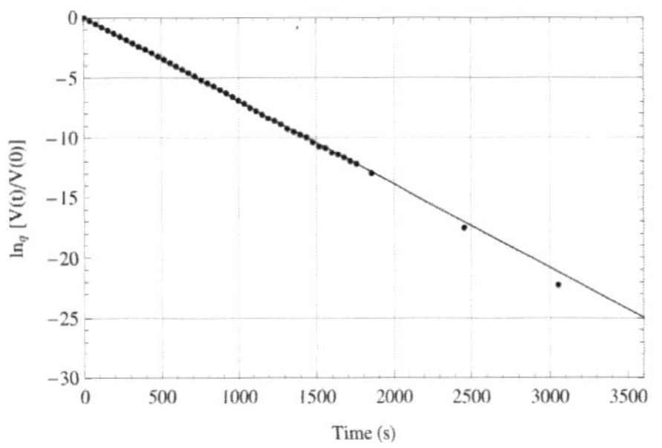

Figure 9. The q-logarithmic plot of the same SPD data in figure 8.

The SPD data is fitted (Figure 8) by using the q-exponential function 4, and plotted with solid line in figure 8. Remarkably the SPD data is well fitted by the q-exponential function with $q = 10.65$, $\tau = 144.2$, and $R^2 = 5.58 \times 10^{-6}$. This is more clear when the same data is expressed in the q-logarithmic plot (figure 9). As can be seen from the equation (7), SPD data in the

q-logarithmic plot becomes a straight line if the q-exponential fitting is well. Indeed this is so as shown in figure 9.

We also checked a possibility of a logarithmic decay, because the shape of logarithmic decay curve looks like a q-exponential curve. We have tried to fit the same SPD data shown in figure 8 with the following fitting function

$$U(t) = \frac{U(0)}{1 + \gamma \ln\left(1 + \frac{t}{\tau}\right)}, \tag{9}$$

where γ and τ are fitting parameters. The broken line in figure 10 is the fitted result for the

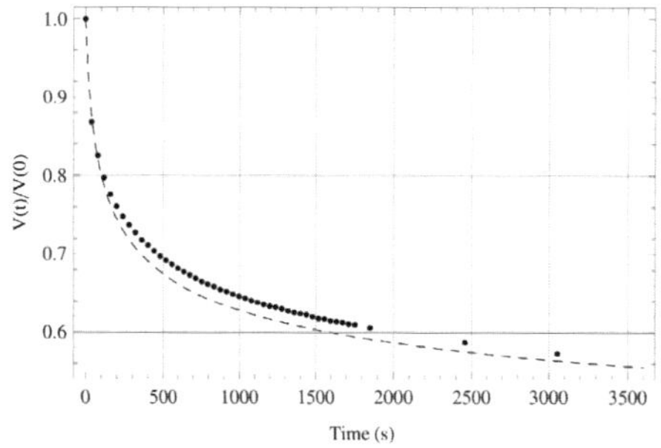

Figure 10. Logarithmic decay of SPD data in figure 8.

same SPD data in figure 8 with the logarithmic decay curve (9). The best fitted parameters are $\gamma = 0.1654$, $\tau = 28.62$, and $R^2 = 2.1 \times 10^{-4}$. As is clear from figure 10, the SPD data of ferrite carrier is not fitted with the logarithmic decay curve (9).

In this way the dielectric materials used for ferrite carrier and electrostatic powder coating are well fitted by the q-exponential function (4) with $q \neq 1$. However, as we will show in the next subsection, the results of nylon powder is different from those of ferrite carrier and electrostatic powder coating.

3.3. Nylon powder for powder coating

Figure 11 is a result of the nylon powders. Evidently $V(t)$ approaches zero as time t passes

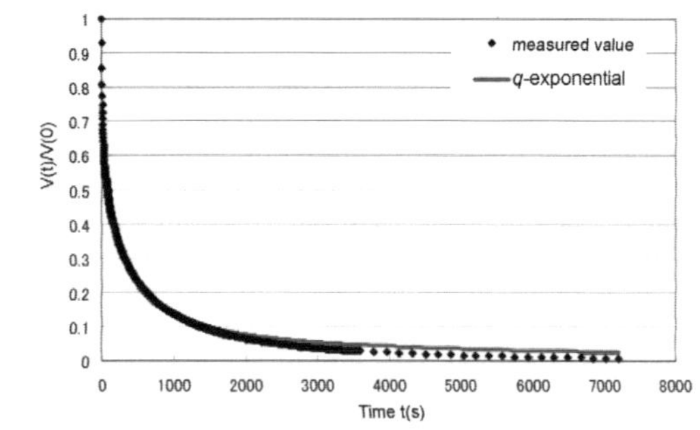

Figure 11. SPD profile of nylon powder. Relative humidity is 52% and temperature is 22.3℃.

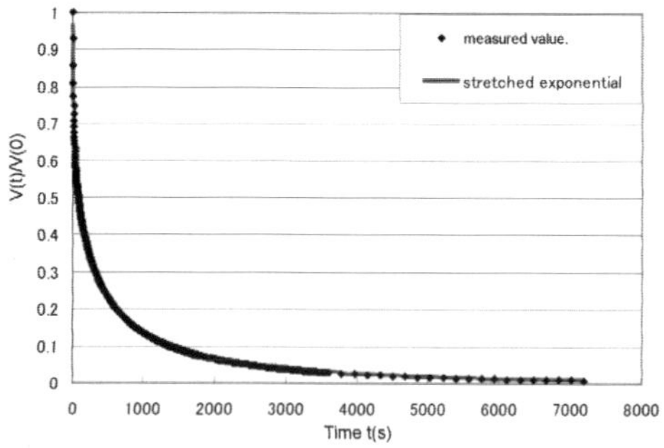

Figure 12. It is fitted by stretched exponential of the same SPD data in figure 11.

due to the low charge retention characteristics of nylon powders. Consequently the SPD data of nylon powder is not fitted by q-exponential function. Instead we confirmed that a stretched exponential function (10) in the following form

$$U(t) = U(0) \exp\left\{ - \left(\frac{t}{\tau}\right)^{\beta} \right\}, \tag{10}$$

is fitted with this data with the fitting parameter $\beta = 0.42$, $\tau = 198$, and $R^2 = 2.83 \times 10^{-5}$. Note that the sum of the square of residuals ($R^2 = 2.83 \times 10^{-5}$) of the stretched exponential fitting (10) is smaller than that ($R^2 = 2.63 \times 10^{-4}$) of the q-exponential fitting (4). Nylon has good electrification, but an increase of moisture content in nylon decreases its electrical insulation performance. A possible reason is therefore that the electrical insulation performance of nylon decreases due to relatively high humidity, then the electric charges may easily escape from the surface of the nylon powders. Consequently the surface potential $V(0)$ soon decays to zero.

4. Concluding remarks

We have studied the surface potential decay of various dielectric material powders: polyvinyl chloride powder; acrylic powder; epoxy powder; polyester powder; ferrite carriers; and nylon powder. Remarkably the most of the SPD profiles are well fitted by Tsallis q-exponential function for the various different samples except nylon powder. A physical reason for having these q-exponential behaviors might be related to some memory effect in the dielectric sample.

5. Acknowledgments

The authors acknowledge to Prof. M. Takeuchi for his advices on the measurements and obtaining some samples. One of the authors (J H) acknowledges the financial supports from research institute for mathematical sciences (RIMS), Kyoto university. We also acknowledge to Prof. C. Tsallis for valuable comments.

6. Reference

[1] Yamada H S and Iguchi K 2008 *Physica* A **387** 1628-1636
[2] Malacarne L C and Mendes R S Lenzi E K 2001 *Phys. Rev.* E **65** 01716
[3] Abe S and Suzuki N 2003 *Phys. Rev* E **67** 016106
[4] Tsallis C, Anjos J C and Borges E P *Phys. Let.* A 2003 **310** 372-376
[5] Malacarne L C and Mendes R S 2000 *Physica* A **286** 319-395
[6] Herous L, Nemamcha M and SAAD S 2006 *J. Eng. Appl. Sci.* **1** 329-334
[7] Tsallis C 1988 *J. Stat. Phys.* **52** 479
[8] Tsallis C *Introduction to Nonextensive Statistical Mechanics: Approaching a Complex World* (Springer, New York 2009).

Journal of Physics: Conference Series **201** (2010) 012011 doi:10.1088/1742-6596/201/1/012011

Molecular dynamics equation designed for realizing arbitrary density: application to sampling method utilizing the Tsallis generalized distribution

Ikuo Fukuda[1] **and Haruki Nakamura**[2]

[1]Computational Science Research Program, RIKEN, 2-1 Hirosawa, Wako, Saitama 351-0198, Japan

[2]Institute for Protein Research, Osaka University, 3-2 Yamadaoka, Suita, Osaka 565-087, Japan

E-mail: ifukuda@riken.jp

Abstract. Several molecular dynamics techniques applying the Tsallis generalized distribution are presented. We have developed a deterministic dynamics to generate an arbitrary smooth density function ρ. It creates a measure-preserving flow with respect to the measure $\rho d\omega$ and realizes the density ρ under the assumption of the ergodicity. It can thus be used to investigate physical systems that obey such distribution density. Using this technique, the Tsallis distribution density based on a full energy function form along with the Tsallis index $q \geq 1$ can be created. From the fact that an effective support of the Tsallis distribution in the phase space is broad, compared with that of the conventional Boltzmann-Gibbs (BG) distribution, and the fact that the corresponding energy-surface deformation does not change energy minimum points, the dynamics enhances the physical state sampling, in particular for a rugged energy surface spanned by a complicated system. Other feature of the Tsallis distribution is that it provides more degree of the nonlinearity, compared with the case of the BG distribution, in the deterministic dynamics equation, which is very useful to effectively gain the ergodicity of the dynamical system constructed according to the scheme. Combining such methods with the reconstruction technique of the BG distribution, we can obtain the information consistent with the BG ensemble and create the corresponding free energy surface. We demonstrate several sampling results obtained from the systems typical for benchmark tests in MD and from biomolecular systems.

1. Introduction

Tsallis' statistics has been extensively investigated [1], and its studies include the fundamental aspect of statistical physics [2] and various applications to physical systems, such as granular matter and turbulence, and to data analysis for network flow and economics [3]. The explanation in view of the Tsallis' statistics are possible for certain physical phenomena that are intractable by the conventional Boltzmann-Gibbs (BG) extensive statistics [4]. Among the properties of this new statistics, we have paid much attention to the fact that the canonical ensemble theory derived from microcanonical ensemble is not unique and that the Tsallis statistics can be derived [5, 6, 7, 8]. Namely, the distribution representing a subsystem in a closed total system is not limited to the BG distribution, but can be realized by the Tsallis distribution. Utilizing the canonical ensemble for molecular dynamics (MD) is necessary in view of the comparison between the MD simulation and the experiment. MD is a method that calculates interactions among microscopic particles composing the physical system and defines a time development of the system, in order to obtain thermal and dynamical properties of the physical system [9]. Thus, the construction of an MD method realizing the Tsallis distribution is very interesting and important.

In contrast to such a pure physical interest regarding the Tsallis distribution itself, there has been a practical approach such that the Tsallis distribution is used as a tool for effective sampling for the BG distribution, on the basis of the traditional statistical mechanics. BG distribution has been intensively investigated also for biological systems, and many studies have revealed that the properties of such systems are well described through the thermodynamical quantities, including the free energy, based on the BG distribution. For example, a stable three-dimensional protein structure is characterized by the lowest free energy state based on the BG distribution among a variety of putative structures of a polypeptide, which has the appropriate amino-acid sequence for an individual protein. However, such system has a complicated energy surface and it is thus hard to obtain an accurate BG distribution. One of the advantages of using the Tsallis distribution for the sampling of physical states is that we can deform the energy surface in a tractable manner. Specifically, an effective support of the Tsallis distribution in the phase space is broad, compared with that of the conventional BG distribution, since the Tsallis distribution obeys the power law with respect to the energy and decreases gradually in the case of the Tsallis index $q > 1$. Furthermore, the energy-surface deformation by the Tsallis distribution does not change the energy minimum points. For these reasons, This distribution enhances to traverse the energy well, without changing the stable states, and to effectively explore the wide region of the phase space. Another advantage of using the distribution is brought when applying it to the equations of motion of MD method. Specifically, we can effectively gain the ergodicity of the dynamical system, which is required to ensure generating the target distribution. This is because the Tsallis distribution provides a greater degree of the nonlinearity in the equations of motion. In addition, owing to the fact that the Tsallis distribution explicitly equips its functional form, we are free from the direct handling of the derivative of the energy density of states, which is required in the flat distribution method such as multicanonical MD.

For the purpose mentioned above, we require constructing an MD method that deterministically generates the Tsallis distribution. In this process, considering to the non-uniquness of the canonical distribution suggested by the studies of the nonextensive statistics, it is preferable to have the ability to generate an arbitrary distribution, rather than a specific equilibrium distribution. Furthermore, in view of the sampling technique, it is flexible not to limit the distribution. Thus, it is useful to construct an equation that realizes an arbitrary given density function, although this is the opposite approach, where we find an underlying density function in the given equations of motion. Regarding such an attempt, we demonstrate the density dynamics in section 2. Applying this method to the Tsallis distribution, we introduce an MD method generating the Tsallis distribution. This MD method is an extension of the MD method that creates the BG distribution (section 3). Against certain physical systems intractable by the conventional methods due to the limited ability to generate the BG distributions, accurate and efficient results are given by the current method owing to the properties of the Tsallis distribution and the equations of motion (section 4).

2. Density dynamics

Let ρ be a density function, i.e., a smooth, positive, and integrable function defined on a domain Ω in $\mathbb{R}^N$. Consider an ordinary differential equation (ODE) having invariant density ρ; i.e., given ρ becomes the density, with respect to Lebesgue measure $d\omega$ on $\mathbb{R}^N$, of an invariant measure for the flow $\{T_t : \Omega \to \Omega | t \in \mathbb{R}\}$ generated by an ODE

$$\dot{\omega} = X(\omega), \tag{1}$$

where $X : \mathbb{R}^N \supset \Omega \to \mathbb{R}^N$ is a smooth, complete vector field. From a generalized Liouville's theorem, it is sufficient that the vector field X satisfies the Liouville equation [10]

$$\operatorname{div} \rho X = 0. \tag{2}$$

It is shown that, by Birkhoff's individual ergodic theorem [11], a time average of any function f (a Borel measurable function on Ω with $\int_\Omega |f|\, \rho d\omega < +\infty$) has a long-time limit value for $\rho d\omega$-almost every initial point ω. Further, if the flow is ergodic with respect to the measure space $(\Omega,\ \mathcal{L}_N \cap \Omega,\ \rho d\omega)$,

$$\lim_{\mathcal{T}\to\infty} \frac{1}{\mathcal{T}} \int_0^{\mathcal{T}} f\left(T_t(\omega)\right) dt = \int_\Omega f \rho d\omega \bigg/ \int_\Omega \rho d\omega \tag{3}$$

holds almost everywhere. Substitution of $f \equiv \chi_A$, the characteristic function for any Borel set $A \subset \Omega$, yields

$$\lim_{\mathcal{T}\to\infty} \frac{1}{\mathcal{T}} \int_0^{\mathcal{T}} \chi_A\left(T_t(\omega)\right) dt = \text{const} \times \int_A \rho(\omega) d\omega \quad \text{(a.e.)}, \tag{4}$$

indicating that the probability density for the realization of point ω is proportional to the given value $\rho(\omega)$. In the sense that an arbitrary density can be realized, we call (ρ, X), a doublet of objective density ρ and a corresponding field X, density dynamics.

A specific form of objective X can be constructed by referring to the Nosé-Hoover (NH) equation [12, 13, 14]. The NH equation is an ODE that can realize the BG distribution, and has been extensively studied including the related ergodic property [15, 16, 17]. Now, the ODE [18] is

$$\begin{aligned} \dot{x}_i &= \mathrm{D}_{p_i}\Theta(\omega), \quad i = 1, \ldots, n, \\ \dot{p}_i &= -\mathrm{D}_{x_i}\Theta(\omega) - \mathrm{D}_\zeta\Theta(\omega)\, p_i, \quad i = 1, \ldots, n, \\ \dot{\zeta} &= \sum_{j=1}^{n} \mathrm{D}_{p_j}\Theta(\omega)\, p_j - n, \end{aligned} \tag{5}$$

with

$$\Theta = -\ln \rho : \Omega \to \mathbb{R}. \tag{6}$$

Here the phase space point is represented by $\omega \equiv (x, p, \zeta)$, where $x \equiv (x_1, \ldots, x_n)$ and $p \equiv (p_1, \ldots, p_n)$ represent coordinates and momenta, respectively, of the physical system, and ζ is an extended variable ($N \equiv 2n + 1$); $\mathrm{D}_{x_i}\Theta(\omega)$, $\mathrm{D}_{p_i}\Theta(\omega)$, and $\mathrm{D}_\zeta\Theta(\omega)$ denote the partial derivative of Θ at ω with respect to x_i, p_i, and ζ, respectively. It can be directly checked that the Liouville equation (2) holds. A certain condition is required for the function Θ to ensure the completeness of X. We assume the fulfillment of this condition as well as the ergodicity.

Here, we have utilized the mathematical structure of invariant measure for the time development of the system and based our formalism on the ergodicity. But, it is, in general, difficult to judge exactly whether or not a given system is ergodic. Even if we can judge it, the system is not necessarily ergodic. When the system is not ergodic, however, ergodic decomposition theorem [19] ensures that on individual ergodic components, which uniquely decompose the phase space, the corresponding ergodicities are valid. If physically meaningful components are produced, it is natural to consider the ergodicity on each invariant set, provided that such a set and the induced invariant measure can be explicitly described.

3. Tsallis dynamics

The method introduced in the previous section is applied to the Tsallis distribution density of the following form [20],

$$\rho_{\mathrm{Tsallis}}(x,p) = [1-(1-q)\beta E(x,p)]^{q/(1-q)}\,, \tag{7}$$

where $E(x,p)$ is a total energy, the sum of kinetic energy $K(p) \equiv \frac{1}{2}\|p\|^2$ and potential energy $U(x)$, which is defined on a domain $D \subset \mathbb{R}^n$. The parameter q is the Tsallis index, for which we treat $q \geq 1$. The limit of the density as $q \to 1$ is $\rho_{\mathrm{BG}}(x,p) \equiv \exp\left[-\beta E(x,p)\right]$, the traditional BG density. In Eq. (7) we have considered the escort probabilities [4, 20], considering the normalized q-expectation value of E, and the renormalized temperature T' with $\beta = 1/k_{\mathrm{B}}T'$. We often shift the origin of the energy value to handle the unknown energy minimum in the applications. It is noted that our method is not restricted to the form given in Eq. (7), but available to utilize the other forms, including the q-exponential function.

To apply the density, Eq. (7), to Eq. (5), we set

$$\rho\left(x,\,p,\,\zeta\right) \equiv \rho_{\mathrm{Tsallis}}(x,\,p)\,\rho_z\left(\zeta\right). \tag{8}$$

Here we have put a general form of the density component with respect to ζ, as $\rho_z\left(\zeta\right)$, which should contribute to meet the condition for ρ on $\Omega \equiv D \times \mathbb{R}^n \times \mathbb{R}$. Then Eq. (5) yields the Tsallis dynamics (TD) [18, 21]:

$$\begin{aligned} \dot{x}_i &= g(x,p)\,p_i, \quad i=1,\ldots,n, \\ \dot{p}_i &= -g(x,p)\,\mathrm{D}U(x) - \tau\left(\zeta\right)\,p_i, \quad i=1,\ldots,n, \\ \dot{\zeta} &= g(x,p)\,\|p\|^2 - nk_{\mathrm{B}}T', \end{aligned} \tag{9}$$

where

$$\begin{aligned} g(x,p) &\equiv \frac{q}{1-(1-q)E(x,p)/k_{\mathrm{B}}T'}, \\ \tau\left(\zeta\right) &\equiv -k_{\mathrm{B}}T'\mathrm{D}\ln\rho_z\left(\zeta\right). \end{aligned}$$

Here we have multiplied the right-hand side of Eq. (9) by constant factor $k_{\mathrm{B}}T'$, so that we get variables x and p with the ordinary physical dimensions (similar procedure is clearly possible in Eq. (5)). Corresponding to the fact that the $q \to 1$ limit for Tsallis density (7) becomes the BG density, substitution of $q=1$ [and $\rho_z\left(\zeta\right) = \exp\left[-(\beta/2Q)\zeta^2\right]$] into TD equation (9), which generates the Tsallis density, yields the NH equation, which generates the BG density. Namely, the extensive limit of the TD is the (generalized) NH equation. As a similar method, that [22] using only the potential energy $U(x)$ (instead of the total energy) for defining the distribution or that [23] targeting the superextensive region ($0<q<1$) has been proposed. The current method first succeeded in generating the Tsallis distribution of the full energetic form in the subextensive region ($q>1$). Note that also a Hamiltonian approach to achieve the Tsallis distribution, in conjunction with the Winkler's method [24], was proposed [25].

Under the conditions stated above and the ergodicity, a long-time average value of physical variable O represented by a function of x and p exists for almost everywhere and equals to the space average in the Tsallis distribution:

$$\begin{aligned}
&\lim_{\tau\to\infty}\frac{1}{\tau}\int_0^{\tau} O\left(x\left(t\right),p(t)\right)dt \\
&= \int_{\Omega} O(x,p)\,\rho(\omega)\,d\omega \Big/ \int_{\Omega}\rho(\omega)d\omega \\
&= \int_{D\times\mathbb{R}^n} O(x,p)\,\rho_{\mathrm{Tsallis}}(x,\,p)\,dxdp \Big/ \int_{D\times\mathbb{R}^n}\rho_{\mathrm{Tsallis}}(x,p)\,dxdp\;.
\end{aligned} \tag{10}$$

We can reconstruct the BG distribution using ODE (9), by applying the re-weighting technique [26]. Under the same assumptions, from Eq. (10) we see

$$\overline{O\,\rho_{\mathrm{BG}}\,/\,\rho_{\mathrm{Tsallis}}}\Big/\overline{\rho_{\mathrm{BG}}\,/\,\rho_{\mathrm{Tsallis}}} = \int_{D\times\mathbb{R}^n} O(x,p)\,\rho_{\mathrm{BG}}(x,\,p)\,dxdp \Big/ \int_{D\times\mathbb{R}^n}\rho_{\mathrm{BG}}\,dxdp \;\;\text{(a.e.)} \tag{11}$$

provided that the space integrals are finite. Therefore, an expectation of O in the BG distribution is obtained, when we evaluate the long-time average of O scaled by the ratio $\rho_{\mathrm{BG}}\,/\,\rho_{\mathrm{Tsallis}}$ and the long-time average of the ratio. In Eq. (11), the temperature T in $\rho_{\mathrm{BG}}(x,p)=\exp\left[-E(x,p)/T\right]$ is not necessary to accord with T' defined in ρ_{Tsallis}, but can be given by any positive value. In addition, during a single process of integrating ODE (9) with a fixed T', we can calculate the expectation values for many temperatures $T=T_1,T_2,....$ This means that many BG distributions with individual temperatures can be realized in a single integration process of the equation of motion. A suitable setting of the parameters in the distribution is not trivial matter. An attempt of the setting for the purpose of the sampling of states is described in Ref. [27].

4. Application of the Tsallis dynamics

As an illustration of the application of TD method to MD, we discuss how the problems in MD can be solved, regarding the 1-dimensional harmonic oscillator (1HO) and 1-dimensional double-well (1DW) systems. Harmonic oscillation becomes an approximation of the small oscillation around the energy minima. However, even for such a simple motion, the correct BG distribution cannot be successfully created, which is a peculiar problem in MD such as in the NH equation. Most evident example is 1HO, in which the trajectory does not fully explore the phase space and show nonergodic behaviours, leading to the failure of generating the BG distribution. This failure would come from an "insufficient complexity" in the equations of motion, due to the fact that the original 1HO Newton equation is a simple (integrable) and small system ("complexity" of the original physical system is one of the factors (but an important factor) to characterize the "complexity" in the equations of motion of MD). To solve this problem, several methods have been proposed, where the number of the extended variables was increased. For example, by increasing arbitrarily the number of thermostats and enforcing their coupling, the Nosé-Hoover chain method [28] gains the complexity in the equations of motions and has improved the ability to obtain ergodicity, even for a simple given system such as 1HO. Another route to gain the complexity is found by the TD method, which uses only one extended variable. That is, by increasing the nonlinearity in the equation of motion by using the "q-deforming" [18] of the NH equation together with a suitable choice of $\rho_z(\zeta)$, TD can generate the BG distribution for 1HO when we recover the BG distribution by the re-weighting [29].

To explore the energy landscape it is necessary to visit other energy minima, by escaping the harmonic-like motion in one energy well, through an unharmonic motion. The most fundamental model of such a motion is that in 1DW. As the energy well is deep, the trajectory is confined over a long time, so that it needs much time to sufficiently sample the phase space, even if the system is ergodic. Many BG dynamics (BGD), i.e., deterministic equation that can generate the BG distribution in a direct way, have been proposed [30, 31, 32]. However, as the accuracy of BGD is increased, the accuracy of the probability $\propto \exp(-\beta E)$ even at a high energy barrier will be increased, so that the simulation requires a long time to escape beyond such barriers. This second problem in MD is emphasized as "increasing the complexity" of the system, in contrast to the first problem stated above. TD solves this problem, using the slow decay feature of the Tsallis distribution with respect to the energy, which means that high energy states can be realized more frequently than those in the BG distribution and so overcoming the energy barriers is enhanced. In systems having such energy surfaces, it is shown [21, 29] that much effective sampling is performed in TD, compared with that in BGD.

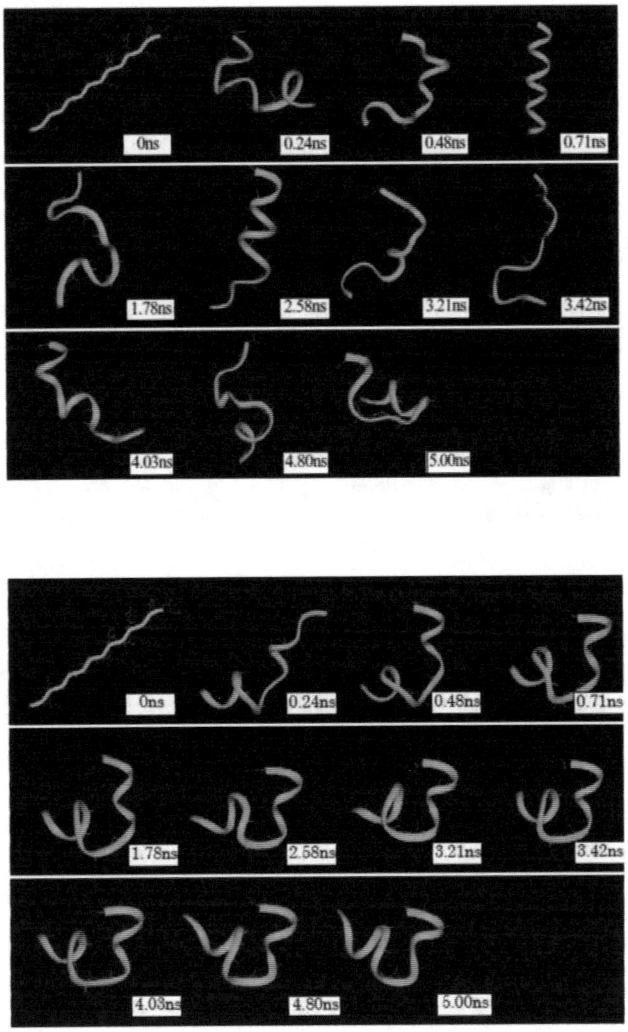

Figure 1. Structural change of main chain of C-peptide obtained in (upper panel) TD simulation and (lower panel) a conventional MD simulation.

As an application of TD to biomolecular system, we show the results of C-peptide, which is a fragment of bovine pancreatic ribonuclease A (PDB id: 1a2w). Figure 1 shows the results of a numerical simulation of 5 ns of the peptide. An extended conformation of the molecule was used as an initial condition. We observe that much effective conformational sampling was performed by the TD (the upper panel of the figure), compared with the results of the conventional MD (BGD at the temperature of 300 K; shown in the lower panel). In the BGD, getting trapped into a local backbone structure is observed, while in the TD a variety of the structures, including the α-helical structure (see e.g., 2.58 ns), which is considered to be the native structure, were sampled much faster. Similar results were confirmed for different initial conditions.

For a development, a molecular-dynamics sampling scheme has been proposed [33] in which any summation of multiple arbitrarily given distributions can be realized deterministically. This method is also based on the density dynamics demonstrated in section 2.

5. Concluding remarks

We have constituted the density dynamics, which is a deterministic method to realize an arbitrary density function. Applying this method to the Tsallis distribution density, we constructed an ODE generating the Tsallis distribution that is described by both the coordinates and the momenta and by the Tsallis index $q \geq 1$, whereas such a construction was impossible in conventional approaches. We demonstrated, via the investigations of 1HO and 1DW systems, that the typical problems in MD were solved by the characteristic features of this dynamics. Application to the peptide system shows that the current method provides efficient results against the case in which the conventional deterministic method unsuccessfully sample the phase space. Other related numerical techniques are shown in Ref. [34], and a recent development relevant to the generalized ensemble method will be discussed in detail elsewhere.

Acknowledgments

This research was supported by Research and Development of the Next-Generation Integrated Simulation of Living Matter, a part of the Development and Use of the Next-Generation Supercomputer Project of the Ministry of Education, Culture, Sports, Science and Technology (MEXT). We thank Masaru Horie for discussions and his valuable help in numerical simulations.

References

[1] Tsallis C 1988 *J. Stat. Phys.* **52** 479
[2] Suyari H 2006 *Physica* A **368** 63
[3] Tsallis C 2009 *Introduction to Nonextensive Statistical Mechanics: Approaching a Complex World* (Berlin: Springer)
[4] Tsallis C 1999 *Braz. J. Phys.* **29** 1; for an updated bibliography, cf. http://tsallis.cat.cbpf.br/biblio.htm.
[5] Plastino A R and Plastino A 1994 *Phys. Lett.* A **193** 140
[6] Abe S and Rajagopal A K 2000 *Phys. Lett.* A **272** 341
[7] Abe S and Rajagopal A K 2000 *J. Phys.* A **33** 8733
[8] Abe S and Rajagopal A K 2001 *Europhys. Lett.* **55** 6
[9] Allen M P and Tildesley D J 1989 *Computer Simulation of Liquids* (Oxford: Oxford Univ. press)
[10] Mañé R 1987 *Ergodic Theory and Differentiable Dynamics* (Berlin: Springer) p 35
[11] Inoue K 1989 *Seminar on Probability* **57** 36
[12] Nosé S 1984 *J. Chem. Phys.* **81** 511
[13] Hoover W G 1985 *Phys. Rev.* A **31** 1695
[14] Nosé S 1991 *Prog. Theor. Phys. Suppl.* **103** 1
[15] Posch H A, Hoover W G and Vesely F J 1986 *Phys. Rev.* A **33** 4253
[16] Cho K and Joannopoulos J D 1992 *Phys. Rev.* A **45** 7089
[17] Di Tolla F D and Ronchetti M 1993 *Phys. Rev.* E **48** 1726
[18] Fukuda I and Nakamura H 2002 *Phys. Rev.* E **65** 026105
[19] Totoki H 1971 *Introduction to ergodic theory* (Tokyo: Kyoritsu)
[20] Tsallis C, Mendes R S and Plastino A R 1998 *Physica* A **261** 534

[21] Fukuda I and Nakamura H 2003 *Chem. Phys. Lett.* **382** 367
[22] Andricioaei I and Straub J E 1997 *J. Chem. Phys.* **107** 9117
[23] Plastino A R and Anteneodo C 1997 *Ann. Phys.* **255** 250
[24] Winkler R G 1992 *Phys. Rev.* A **45** 2250
[25] Andrade J S Jr., Almeida M P, Moreira A A and Farias G A 2002 *Phys. Rev.* E **65** 036121
[26] Ferrenberg A M and Swendsen R H 1988 *Phys. Rev. Lett.* **61** 2635
[27] Fukuda I, Horie M and Nakamura H 2005 *Chem. Phys. Lett.* **405** 364
[28] Martyna G J, Klein M L and Tuckerman M 1992 *J. Chem. Phys.* **97** 2635
[29] Fukuda I and Nakamura H 2004 *J. Phys. Chem.* B **108** 4162
[30] Hoover W G and Holian B L 1996 *Phys. Lett.* A **211** 253; and the references therein
[31] Liu Y and Tuckerman M E 2000 *J. Chem. Phys.* **112** 1685
[32] Sergi A and Ferrario M 2001 P*hys. Rev. E* **64** 056125
[33] Fukuda I and Nakamura H 2005 *Phys. Rev.* E **71** 046708
[34] Fukuda I and Nakamura H 2006 *Phys. Rev.* E **73** 026703

Mathematical Aspects of Generalized Entropies and their Applications IOP Publishing
Journal of Physics: Conference Series **201** (2010) 012012 doi:10.1088/1742-6596/201/1/012012

A dually flat structure on the space of escort distributions

Atsumi Ohara
Department of Systems Science, Graduate School of Engineering Science, Osaka University, Toyonaka, Osaka 560-8531 Japan
E-mail: ohara@sys.es.osaka-u.ac.jp

Hiroshi Matsuzoe
Department of Computer Science and Engineering, Graduate School of Engineering, Nagoya Institute of Technology, Gokiso-cho, Showa-ku, Nagoya, Aichi 466-8555 Japan
E-mail: matsuzoe@nitech.ac.jp

Shun-ichi Amari
Riken Brain Science Institute Wako-shi Hirosawa 2-1, Saitama 351-0198 Japan
E-mail: amari@brain.riken.jp

Abstract. This note studies geometrical structure of the manifold of escort probability distributions and proves that the resultant geometry is dually flat in the sense of information geometry. We use a conformal transformation that flattens the alpha-geometry of the space of the discrete probability distributions in order to realize escort probabilities in the framework of affine differential geometry. Dual pairs of potential functions and affine coordinate systems on the manifold are derived, and the associated canonical divergence is shown to be conformal to the alpha-divergence.

1. Introduction

Escort probabilities are naturally induced from the studies of multifractals [3] and non-extensive entropy (Tsallis entropy) [20, 21] to play an important but mysterious role. The purpose of this note is to investigate the escort probability from viewpoints of information geometry [1, 2] and affine differential geometry [12].

Recently, it is reported that -geometry, which is an information geometric structure with constant curvature, has a close relation with the Tsallis entropy [13, 14]. The main feature of the -geometry consists of a Riemannian metric together with a one-parameter family of dual affine connections, called the -connections.

We show that the manifold of escort probability distributions is dually flat by considering conformal transformations that flatten the -geometry on the manifold of usual probability distributions. On the resultant manifold, escort probabilities consist of an affine coordinate system. See also [17] for another type of flattening a curved dual manifold by a conformal transformation.

This gives us a clear geometrical interpretation of the escort probability, and simultaneously, produces its new obscure links to conformality and projectivity. Due to these two geometrical concepts, however, the obtained dually flat structure inherits several properties of the -geometry. Further, we show a pair of potential functions and derive dual affine coordinate systems for the structure.

Section 2 is devoted to preliminaries for the -geometry in the light of the affine differential geometry. In section 3, we consider the conformal transformations and discuss properties of the obtained dually flat structure.

In the sequel, all the affine connections are assumed torsion-free. The exponent q is fixed to be $(1 \quad)/2$ for $\quad 1 < \quad < 1$.

2. Preliminaries

We briefly introduce the -geometry via the affine differential geometry. See for details [13, 14]. Let $\mathcal{S}^n$ denote the n-dimensional probability simplex, i.e.,

$$\mathcal{S}^n := \left\{ \boldsymbol{p} = (p_i) \quad p_i > 0, \sum_{i=1}^{n+1} p_i = 1 \right\}, \tag{1}$$

and $p_i, i = 1, \quad , n+1$ denote probabilities of $n+1$ states. We introduce the -geometric structure on $\mathcal{S}^n$. Let $\{\partial_i\}, i = 1, \quad , n$ be natural basis tangent vector fields on $\mathcal{S}^n$ defined by

$$\partial_i := \frac{\partial}{\partial p_i} \quad \frac{\partial}{\partial p_{n+1}}, \quad i = 1, \quad , n, \tag{2}$$

where $p_{n+1} = 1 \quad \sum_{i=1}^n p_i$. Now we define a Riemannian metric g on $\mathcal{S}^n$ by the Fisher information matrix,

$$\begin{aligned} g_{ij}(\boldsymbol{p}) &:= g(\partial_i, \partial_j) = \frac{1}{p_i} \quad _{ij} + \frac{1}{p_{n+1}} \\ &= \sum_{k=1}^{n+1} p_k (\partial_i \log p_k)(\partial_j \log p_k), \quad i, j = 1, \quad , n. \end{aligned} \tag{3}$$

Further, define an affine connection $\nabla^{(\ )}$ called the *-connection*, which is represented in its coefficients by

$$\Gamma_{ij}^{(\ )k}(\boldsymbol{p}) = \frac{1 + \quad}{2} \quad \frac{1}{p_k} \quad _{ij}^{k} + p_k g_{ij} \quad , \quad i, j, k = 1, \quad , n, \tag{4}$$

where $\ _{ij}^{k}$ is equal to one if $i = j = k$ and zero otherwise. Then we have the -covariant derivative $\nabla^{(\ )}$, which gives

$$\nabla_{\partial_i}^{(\ )} \partial_j = \sum_{k=1}^{n} \Gamma_{ij}^{(\ )k} \partial_k,$$

when it is applied to the vector fields ∂_i and ∂_j.

There are two specific features for the -geometry on $\mathcal{S}^n$ defined in such a way. First, the triple $(\mathcal{S}^n, g, \nabla^{(\ )})$ is a *statistical manifold* [8] (See the appendix for its definition), i.e., we can confirm that the following relation holds:

$$X g(Y, Z) = g(\nabla_X^{(\ )} Y, Z) + g(Y, \nabla_X^{(\quad)} Z), \quad X, Y, Z \in \mathcal{X}(\mathcal{S}^n), \tag{5}$$

where $\mathcal{X}(\mathcal{S}^n)$ denotes the set of all tangent vector fields on $\mathcal{S}^n$. Two statistical manifolds $(\mathcal{S}^n, g, \nabla^{(\ )})$ and $(\mathcal{S}^n, g, \nabla^{(\quad)})$ are said *mutually dual.*

The other is that $(\mathcal{S}^n, g, \nabla^{(\ )})$ is a manifold with constant curvature $\ = (1 \quad ^2)/4$, i.e.,

$$R^{(\ )}(X, Y)Z = \ \{g(Y, Z)X \quad g(X, Z)Y\},$$

where $R^{(\ )}$ is the curvature tensor with respect to $\nabla^{(\ )}$. From this property the well-known nonadditive formula of the Tsallis entropy can be derived [13].

In [14] we have discussed the -geometry on $\mathcal{S}^n$ from a viewpoint of the *affine differential geometry* [12]. Consider the immersion f of $\mathcal{S}^n$ into $\mathbf{R}_+^{n+1}$ by

$$f : \boldsymbol{p} = (p_i) \mapsto \boldsymbol{x} = (x^i) = (L^{(\ )}(p_i)), \quad i = 1, \quad , n+1, \tag{6}$$

where (x^i) is the canonical flat coordinate system of $\mathbf{R}^{n+1}$ and the function $L^{(\ )}$ is defined by

$$L^{(\ )}(t) := \frac{2}{1 \quad } t^{(1 \quad)/2} = \frac{1}{q} t^q.$$

Note that $f(\mathcal{S}^n)$ is a level hypersurface in the ambient space $\mathbf{R}_+^{n+1}$ represented by $\Psi(\boldsymbol{x}) = 2/(1 + \ )$, where

$$\Psi(\boldsymbol{x}) := \frac{2}{\ + 1} \sum_{i=1}^{n+1} \frac{1 \quad }{2} x^i \quad ^{2/(1 \quad)} = \frac{1}{1 \quad q} \sum_{i=1}^{n+1} \quad q x^i \quad ^{1/q}. \tag{7}$$

We choose a transversal vector on the level hypersurface by

$$:= \sum_{i=1}^{n+1} \ ^i \frac{\partial}{\partial x^i}, \quad ^i = \quad q(1 \quad q)x^i = \quad x^i. \tag{8}$$

Then we can confirm that the affine immersion $(f, \ )$ realizes the -geometry on $\mathcal{S}^n$ [14]. Hence, it would be possible to develop theory of the -geometry and Tsallis statistics with ideas of the affine differential geometry [11].

Further, the *escort probability* [3] naturally appears in this setup. The escort probability $\boldsymbol{P} = (P_i)$ associated with $\boldsymbol{p} = (p_i)$ is the normalized version of $(p_i)^q$, and is defined by

$$P_i(\boldsymbol{p}) := \frac{(p_i)^q}{\sum_{i=1}^{n+1}(p_i)^q} = \frac{x^i}{Z_q}, \quad Z_q(\boldsymbol{p}) := \sum_{i=1}^{n+1} x^i(\boldsymbol{p}), \ \boldsymbol{x}(\boldsymbol{p}) \in f(\mathcal{S}^n). \tag{9}$$

Hence, the simplex $\mathcal{E}^n$ in the ambient space $\mathbf{R}_+^{n+1}$, i.e.,

$$\mathcal{E}^n := \left\{ \boldsymbol{x} = (x^i) \quad \sum_{i=1}^{n+1} x^i = 1, \ x^i > 0 \right\}$$

represents the set of escort distributions $\boldsymbol{P}$.

Note that the element $\boldsymbol{x}^* = (x_i^*)$ in the dual space of $\mathbf{R}^{n+1}$ defined by

$$x_i^*(\boldsymbol{p}) := L^{(\quad)}(p_i) = \frac{1}{1 \quad q}(p_i)^{1 \quad q}, \quad i = 1, \quad , n+1,$$

meets

$$x_i^*(\boldsymbol{p}) = \frac{\partial \Psi}{\partial x^i}(\boldsymbol{x}(\boldsymbol{p})).$$

Hence, it satisfies [14]

$$\sum_{i=1}^{n+1} \ ^i(\boldsymbol{p}) x_i^*(\boldsymbol{p}) = 1, \quad \sum_{i=1}^{n+1} x_i^*(\boldsymbol{p}) X^i = 0, \tag{10}$$

for an arbitrary vector $X = \sum_{i=1}^{n+1} X^i \partial/\partial x^i$ at $\boldsymbol{x}(\boldsymbol{p})$ tangent to $f(\mathcal{S}^n)$. Thus, $\ \boldsymbol{x}^*(\boldsymbol{p})$ can be interpreted as the *conormal map* [12].

3. A conformally and projectively flat geometric structure and escort probabilities

In this section we consider a conformal and projective transformation [5, 6, 7, 10] of the -geometry to introduce a dually flat one. This flattening of the -geometry conserves some of its properties. The escort probabilities (P_i) are found to represent one of mutually dual affine coordinate systems in the induced geometry.

Let us define a function on $\mathcal{S}^n$ by

$$(\boldsymbol{p}) := \frac{1}{Z_q} = \frac{1}{\sum_{i=1}^{n+1} L^{(\ )}(p_i)},$$

which depends on . Then, from (9) $\mathcal{E}^n$ is regarded as the image of $\mathcal{S}^n$ for another immersion $\tilde{f} := \ f$, i.e.,

$$\tilde{f} : \mathcal{S}^n \ni (p_i) \mapsto (P_i) \in \mathcal{E}^n,$$

and $(P_i)_{i=1}^n$ is interpreted as a coordinate system of $\mathcal{S}^n$. It would be a natural way to introduce geometric structure on $\mathcal{E}^n$ (and hence on $\mathcal{S}^n$) via the affine immersion $(\tilde{f}, \tilde{\ })$ by taking a suitable transversal vector $\tilde{\ }$, similarly to the case of the -geometry mentioned above. Since $\mathcal{E}^n$ is a part of a hyperplane in $\mathbf{R}^{n+1}$, the canonical affine connection of $\mathbf{R}^{n+1}$ induces a flat connection, denoted by $D^{(\mathrm{E})}$, on $\mathcal{E}^n$. However, for the same reason, we cannot define a Riemannian metric in this way[1] regardless of any choice of the transversal vector $\tilde{\ }$.

The idea we adopt here is to define a Riemannian metric by utilizing a property of $(\mathcal{S}^n, g, \nabla^{(\ )})$ called 1-*conformal flatness*. Based on the results proved by Kurose [5, 6], we conclude that the manifold $(\mathcal{S}^n, g, \nabla^{(\ )})$ is 1-conformally flat (See the appendix for its definition) because it is a statistical manifold with constant curvature.

Actually, let ∇^* be the flat connection[2] on $\mathcal{S}^n$ pulled back by $\tilde{f}$ from $(\mathcal{E}^n, D^{(\mathrm{E})})$. Then, we can prove that $\nabla^{(\ )}$ and ∇^* are *projectively equivalent*, i.e., it holds that

$$\nabla^*_X Y = \nabla^{(\ )}_X Y + d(\ln \ )(Y)X + d(\ln \ )(X)Y, \quad X, Y \in \mathcal{X}(\mathcal{S}^n). \tag{11}$$

Hence, if we define another Riemannian metric h on $\mathcal{S}^n$ by

$$h(X,Y) := \ g(X,Y), \quad X, Y \in \mathcal{X}(\mathcal{S}^n), \tag{12}$$

then, $(\mathcal{S}^n, g, \nabla^{(\ )})$ is 1-conformally equivalent to $(\mathcal{S}^n, h, \nabla^*)$ having a flat connection ∇^*. Further, the manifold $(\mathcal{S}^n, h, \nabla^*)$ can be proved to be a statistical manifold.

Using the conormal map $\boldsymbol{x}^*(\boldsymbol{p})$, we can define the *-divergence* as a *contrast function*[3] on $(\mathcal{S}^n, g, \nabla^{(\ )})$ as follows [6]:

$$\begin{aligned} D^{(\ )}(\boldsymbol{p}, \boldsymbol{r}) &= \sum_{i=1}^{n+1} x_i^*(\boldsymbol{r})(x^i(\boldsymbol{p}) \quad x^i(\boldsymbol{r})) \\ &= \langle \ \boldsymbol{x}^*(\boldsymbol{r}), \boldsymbol{x}(\boldsymbol{p}) \quad \boldsymbol{x}(\boldsymbol{r})\rangle = \frac{1}{\ } \quad \langle \boldsymbol{x}^*(\boldsymbol{r}), \boldsymbol{x}(\boldsymbol{p})\rangle. \end{aligned}$$

The statistical manifolds $(\mathcal{S}^n, g, \nabla^{(\ \ )})$ and $(\mathcal{S}^n, g, \nabla^{(\ )})$ are dual in the sense of (5). Further, it is known [2] that there exists the unique dual affine flat connection ∇ on $\mathcal{S}^n$ with respect to (h, ∇^*). Then, $(\mathcal{S}^n, g, \nabla^{(\ \ )})$ is proved 1-conformally equivalent to $(\mathcal{S}^n, h, \nabla)$ [6]. This fact

[1] In the affine differential geometry, a Riemannian metric is realized as the affine fundamental form of an affine immersion [12].

[2] For the sake of notational consistency with the existing literature, e.g., [1, 2], we first define ∇^*, and later ∇ as the dual of ∇^*.

[3] A divergence that is compatible with statistical manifold structure in the sense of Eguchi [4].

directly implies [6] that a contrast function ρ of $(\mathcal{S}^n, h, \nabla)$ is given by scaling $D^{(\quad)}$ that is a contrast function of $(\mathcal{S}^n, g, \nabla^{(\quad)})$, i.e.,

$$\begin{aligned}\rho(\boldsymbol{p}, \boldsymbol{r}) &= \quad(\boldsymbol{r})D^{(\quad)}(\boldsymbol{p}, \boldsymbol{r}) = \frac{1}{Z_q(\boldsymbol{r})}D^{(\quad)}(\boldsymbol{p}, \boldsymbol{r}) \\ &= \frac{1}{Z_q(\boldsymbol{r})}\langle\ \boldsymbol{x}(\boldsymbol{r}), \boldsymbol{x}^*(\boldsymbol{p}) \quad \boldsymbol{x}^*(\boldsymbol{r})\rangle = \langle\ \boldsymbol{P}(\boldsymbol{r}), \boldsymbol{x}^*(\boldsymbol{p}) \quad \boldsymbol{x}^*(\boldsymbol{r})\rangle. \end{aligned} \tag{13}$$

We shall call ρ a *conformal divergence.*

Since $(\mathcal{S}^n, h, \nabla, \nabla^*)$ is a dually flat structure, it is equipped with mutually dual affine coordinate systems (θ^i) and $(\ _i)$, $i = 1, \quad, n$, a potential function $\psi(\boldsymbol{\theta})$ and its conjugate $\psi^*(\boldsymbol{\eta})$ [1, 2] satisfying

$$_i = \frac{\partial\psi}{\partial\theta^i}, \quad \theta^i = \frac{\partial\psi^*}{\partial\ _i}.$$

Further, a contrast function is written in the form of the *canonical divergence* [2], i.e.,

$$\rho(\boldsymbol{p}, \boldsymbol{r}) = \psi(\boldsymbol{\theta}(\boldsymbol{p})) + \psi^*(\boldsymbol{\eta}(\boldsymbol{r})) \quad \sum_{i=1}^{n}\theta^i(\boldsymbol{p})\ _i(\boldsymbol{r}). \tag{14}$$

From these facts, we see that the following holds:

Proposition 1 *For the dually flat structure* $(\mathcal{S}^n, h, \nabla, \nabla^*)$ *defined via* $\quad 1$*-conformal transformation from* $(\mathcal{S}^n, g, \nabla^{(\ )}, \nabla^{(\quad)})$*, the associated potential functions and dually flat affine coordinate systems are represented as follows:*

$$\begin{aligned} &_i(\boldsymbol{p}) = P_i(\boldsymbol{p}), && \theta^i(\boldsymbol{p}) = x_i^*(\boldsymbol{p}) \quad x_{n+1}^*(\boldsymbol{p}), \quad i = 1, \quad, n \\ \psi^*(\boldsymbol{\eta}(\boldsymbol{p})) &= \frac{1}{\ }(\ (\boldsymbol{p}) \quad q), && \psi(\boldsymbol{\theta}(\boldsymbol{p})) = \quad \ln_q(p_{n+1}), \end{aligned}$$

where $\ln_q(t) := (t^{1\quad q} \quad 1)/(1 \quad q)$*, and* $\ = (1 \quad ^2)/4 = q(1 \quad q)$ *is the scalar curvature. Here, the coordinate systems* (θ^i) *and* $(\ _i)$ *are* ∇*- and* ∇^**-affine, respectively.*

Proof) We have only to check that the canonical divergence coincides with a contrast function of $(\mathcal{S}^n, h, \nabla)$. Substitute the potential functions and dual affine coordinates in the statement to the right-hand side of (14), then we see that it coincides with $\rho(\boldsymbol{p}, \boldsymbol{r})$ in (13). Q.E.D.

Corollary 1 *The escort probabilities* $P_i, i = 1, \quad, n$ *are canonical affine coordinates of the flat affine connection* ∇^* *on* $\mathcal{S}^n$.

There are three remarks. First, since it holds that

$$\ln_q(p_i) = x_i^*(\boldsymbol{p}) \quad \frac{1}{1 \quad q},$$

we can alternatively represent

$$\theta^i(\boldsymbol{p}) = \ln_q(p_i) \quad \ln_q(p_{n+1}).$$

Further, note that the conformal factor $\quad$ is alternatively represented as follows:

$$(\boldsymbol{p}) = \frac{1}{Z_q(\boldsymbol{p})} = \frac{q}{\sum_{i=1}^{n+1}(p_i)^q} = \frac{q}{(\exp_q(S_q(\boldsymbol{p})))^{1\quad q}} = \quad \ln_q \quad \frac{1}{\exp_q(S_q)}\Bigg) + q, \tag{15}$$

where

$$S_q(\boldsymbol{p}) := \frac{\sum_{i=1}^{n+1}(p_i)^q \quad 1}{1 \quad q}, \quad \exp_q(t) := [1 + (1 \quad q)t]_+^{1/(1 \quad q)}.$$

Substituting (15) into the expression of ψ^*, we have

$$\psi^* = \ln_q \quad \frac{1}{\exp_q(S_q)}\Bigg).$$

We easily see that the expressions of θ^i and ψ^* here and those of $\ _i$ and ψ in the proposition recover the standard ones [1, 2] when $q \to 1$. Note that $\ \psi^*$ coincides with the entropy studied in [9, 18, 22] and refered to as the *normalized Tsallis entropy*. The conformal factor (or scaling factor) often appears in the study of the q-analysis.

Next, similarly to the above conformal transformation of $(\mathcal{S}^n, g, \nabla^{(\ )})$, we can define another one for $(\mathcal{S}^n, g, \nabla^{(\ \ )})$ with a conformal factor

$$'(\boldsymbol{p}) := \frac{1}{\sum_{i=1}^{n+1} L^{(\ \ )}(p_i)},$$

and construct another dually flat structure $(\mathcal{S}^n, h' = \ 'g, \nabla', \nabla'^*)$. Hence, the relations among them are shown in the Figure 1.

	$(\mathcal{S}^n, h', \nabla')$	$\overset{\text{dual}}{\longleftrightarrow}$	$(\mathcal{S}^n, h', \nabla'^*)$	
1-conformally equivalent	$\updownarrow$		$\updownarrow$	1-conformally equivalent
	$(\mathcal{S}^n, g, \nabla^{(\ )})$	$\overset{\text{dual}}{\longleftrightarrow}$	$(\mathcal{S}^n, g, \nabla^{(\ \ )})$	
1-conformally equivalent	$\updownarrow$		$\updownarrow$	1-conformally equivalent
	$(\mathcal{S}^n, h, \nabla^*)$	$\overset{\text{dual}}{\longleftrightarrow}$	$(\mathcal{S}^n, h, \nabla)$	

Figure 1. Relations among geometries

Finally, because of the projective equivalence (11), a submanifold in $\mathcal{S}^n$ is $\nabla^{(\ )}$-autoparallel if and only if it is ∇-autoparallel. In particular, distributions constrained with the normalized q-expectations (escort averages) [21] consist of a simultaneously $\nabla^{(\ )}$- and ∇-autoparallel submanifold in $\mathcal{S}^n$.

4. Concluding remarks

We have considered 1-conformal transformations of the -geometry and obtained dually flat structure $(\mathcal{S}^n, h, \nabla, \nabla^*)$. Further the potential functions and dually flat coordinate systems associated with the structure have been derived. We see that the escort probability naturally appears to play an important role.

From a viewpoint of contrast functions, the geometric structure compatible to the Kullback-Leibler divergence is $(\mathcal{S}^n, g, \nabla^{(1)}, \nabla^{(\ 1)})$, where g is the Fisher information and $\nabla^{(\pm 1)}$ are respectively the *e-connection* and the *m-connection*. Similarly, the -divergence (or the Tsallis relative entropy), and the conformal divergence ρ in this note correspond to $(\mathcal{S}^n, g, \nabla^{(\ )}, \nabla^{(\ \ )})$ and $(\mathcal{S}^n, h, \nabla, \nabla^*)$, respectively. They are summarized in Figure 2. The physical meaning or essence underlying these transformations would be interesting and significant.

We hope that arguments in this note might be helpful to more profound investigations and understandings of this area [15, 16, 19].

KL divergence		-divergence		conformal divergence
$(\mathcal{S}^n, g, \nabla^{(1)}, \nabla^{(\ 1)})$	$\longleftrightarrow$	$(\mathcal{S}^n, g, \nabla^{(\ )}, \nabla^{(\ \ )})$	$\longleftrightarrow$	$(\mathcal{S}^n, h, \nabla, \nabla^*)$, $(\mathcal{S}^n, h', \nabla', \nabla'^*)$
dually flat		constant curvature		dually flat

Figure 2. transformations of dualistic structures

Acknowledgments
We would thank the anonymous referee who has pointed out several errors in the manuscript and important literature.

Appendix: Statistical manifold and -conformally equivalence

For details of this appendix see [8, 5, 6, 7, 10]. For a torsion-free affine connection ∇ and a pseudo Riemannian metric g on a manifold $\mathcal{M}$, the triple $(\mathcal{M}, g, \nabla)$ is called a *statistical manifold* if it admits another torsion-free connection ∇^* satisfying

$$Xg(Y,Z) = g(\nabla_X Y, Z) + g(Y, \nabla^*_X Z) \tag{A.1}$$

for arbitrary X, Y and Z in $\mathcal{X}(\mathcal{M})$, where $\mathcal{X}(\mathcal{M})$ is the set of all tangent vector fields on $\mathcal{M}$. It is known that $(\mathcal{M}, g, \nabla)$ is a statistical manifold if and only if ∇g is symmetric. We call ∇ and ∇^* *duals* of each other with respect to g, and $(\mathcal{M}, g, \nabla^*)$ is said the dual statistical manifold of $(\mathcal{M}, g, \nabla)$. The triple of a Riemannian metric and a pair of dual connections (g, ∇, ∇^*) satisfying (A.1) is called a *dualistic structure* on $\mathcal{M}$.

For $\ \in \mathbf{R}$, statistical manifolds $(\mathcal{M}, g, \nabla)$ and $(\mathcal{M}, g', \nabla')$ are said to be *-conformally equivalent* if there exists a function ϕ on $\mathcal{M}$ such that

$$\begin{aligned} g'(X,Y) &= e^{\phi} g(X,Y), \\ g(\nabla'_X Y, Z) &= g(\nabla_X Y, Z) \quad \frac{1+\ }{2} d\phi(Z) g(X,Y) + \frac{1\ \ }{2}\{d\phi(X) g(Y,Z) + d\phi(Y) g(X,Z)\}. \end{aligned}$$

Statistical manifolds $(\mathcal{M}, g, \nabla)$ and $(\mathcal{M}, g', \nabla')$ are -conformally equivalent if and only if $(\mathcal{M}, g, \nabla^*)$ and $(\mathcal{M}, g, \nabla'^*)$ are -conformally equivalent.

A statistical manifold $(\mathcal{M}, g, \nabla)$ is called *-conformally flat* if it is locally -conformally equivalent to a flat statistical manifold. Note that 1-conformal equivalence implies projective equivalence. A statistical manifold of dimension greater than three has constant curvature if and only if it is 1-conformally flat. If $(\mathcal{M}, g, \nabla)$ and $(\mathcal{M}, g', \nabla')$ are 1-conformally equivalent, a contrast function ρ' of $(\mathcal{M}, g', \nabla')$ is given by

$$\rho'(p,q) = e^{\phi(q)} \rho(p,q),$$

where ρ is a contrast function of $(\mathcal{M}, g, \nabla)$.

References

[1] Amari S-I 1985 *Differential-Geometrical Methods in Statistics*, (Berlin Springer-Verlag)
[2] Amari S-I and Nagaoka H 2000 *Methods of Information Geometry*, (Rhode Island: AMS&Oxford).
[3] Beck C and Schlögl F 1993 *Thermodynamics of Chaotic Systems* (Cambridge: Cambridge University Press).
[4] Eguchi S 1992 Geometry of minimum contrast *Hiroshima Math. J.* **23** 631–47
[5] Kurose T 1990 Dual Connections and Affine Geometry *Math. Z.* **203** 115-21
[6] Kurose T 1994 On the Divergence of 1-conformally Flat Statistical Manifolds, *Tôhoku Math. J.* **46** 427-33
[7] Kurose T 2002 Conformal-projective geometry of statistical manifolds. *Interdiscip. Inform. Sci.* **8** 89–100

[8] Lauritzen S L 1987 Statistical manifolds S.Amari *et al* eds. *Differential geometry in statistical inference* 163–216 (Hayward: Institute of Mathematical Statistics)
[9] Landsberg P T and Vedral V 1998 Distributions and channel capacities in generalized statistical mechanics *Phys. Lett. A* **247** 211–7
[10] Matsuzoe H 1999 Geometry of contrast functions and conformal geometry *Hiroshima Math. J.* **29** 175–91
[11] Matsuzoe H, Takeuchi J and Amari S-I 2006 Equiaffine structures on statistical manifolds and Bayesian statistics *Differential Geom. Appl.* **24** 567–78
[12] Nomizu K and Sasaki T 1993 *Affine Differential Geometry* (Cambridge: Cambridge University Press)
[13] Ohara A 2007 Geometry of distributions associated with Tsallis statistics and properties of relative entropy minimization *Phys. Lett. A* **370** 184–93
[14] Ohara A 2009 Geometric study for the Legendre duality of generalized entropies and its application to the porous medium equation *Euro. Phys. J. B* **70** 15–28
[15] Ojima I 2004 Temperature as order parameter of broken scale invariance *Publ. RIMS (Kyoto Univ.)* **40** 731-56
[16] Ojima I 2009 Meaning of Non-Extensive Entropies in Micro-Macro Duality *to be included in the present proceedings*
[17] Okamoto I, Amari S and Takeuchi K 1991 Asymptotic theory of sequential estimation procedures for curved exponential families *Annals of Statistics* **19** 961-81
[18] Rajagopal A K and Abe S 1999 Implications of Form Invariance to the Structure of Nonextensive Entropies *Phys. Rev. Lett.* **83** 1711-4
[19] Tanaka M 2009 Meaning of an escort distribution and τ-transformation *to be included in the present proceedings*
[20] Tsallis C 1988 Possible generalization of Boltzmann-Gibbs statistics *J. Stat. Phys.* **52**, 479-87
[21] Tsallis C 2009 *Introduction to Nonextensive Statistical Mechanics: Approaching a Complex World* (Berlin/Heidelberg: Springer)
[22] Wada T and Scarfone A M 2005 Connections between Tsallis' formalisms employing the standard linear average energy and ones employing the normalized q-average energy *Phys. Lett. A* **335** 351–62

Journal of Physics: Conference Series **201** (2010) 012013 doi:10.1088/1742-6596/201/1/012013

Towards operational interpretations of generalized entropies

Flemming Topsøe

University of Copenhagen, Department of Mathematical Sciences, Universitetsparken 5, 2100 Copenhagen, Denmark

E-mail: topsoe@math.ku.dk

Abstract. The driving force behind our study has been to overcome the difficulties you encounter when you try to extend the clear and convincing operational interpretations of classical Boltzmann-Gibbs-Shannon entropy to other notions, especially to generalized entropies as proposed by Tsallis. Our approach is philosophical, based on speculations regarding the interplay between *truth, belief* and *knowledge*. The main result demonstrates that, accepting philosophically motivated assumptions, the only possible measures of entropy are those suggested by Tsallis – which, as we know, include classical entropy. This result constitutes, so it seems, a more transparent interpretation of entropy than previously available. However, further research to clarify the assumptions is still needed. Our study points to the thesis that one should never consider the notion of entropy in isolation – in order to enable a rich and technically smooth study, further concepts, such as *divergence*, *score functions* and *descriptors* or *controls* should be included in the discussion. This will clarify the distinction between *Nature* and *Observer* and facilitate a game theoretical discussion. The usefulness of this distinction and the subsequent exploitation of game theoretical results – such as those connected with the notion of *Nash equilibrium* – is demonstrated by a discussion of the *Maximum Entropy Principle.*

1. Introduction

Our aim is to provide transparent operational interpretations of generalized notions of entropy, especially of *Tsallis entropy*, cf. [1] and [2]. Our approach depends on abstract, philosophical considerations centred around notions of *truth, belief* and *knowledge* and their possible interplay. In Section 2 we introduce the main abstract notions. They are not necessarily tied to probabilistic considerations. However, for the present exposition, we tone down a bit the very abstract discussion.

In Section 3, we introduce the probabilistic models we shall work with. We only consider discrete models. For these models, we identify the most natural or acceptable forms of interplay between truth, belief and knowledge. Key notions are related to *interaction* and *description.*

Quantitative reasoning is enabled by the introduction of *score functions.* This type of function is known from statistical decision theory. Here, it determines the *effort* needed by an observer in order to gain knowledge. A variational principle is introduced which is related to the *fundamental inequality of information theory.* The notions introduced are needed for the formulation of two main results, Theorems 1 and 2. In particular, Theorem 2 is presented as an operational interpretation of Tsallis entropy. It singles out Tsallis entropy among other possibilities and may thus be taken to support the view that these notions occupy a unique position in statistical physics.

In Section 4 we introduce concepts closely related to *experiments* and *observation*. The further study depends on game theoretical considerations and this is taken up in Section 5. Two notions of equilibrium are introduced and their relation is established in Theorem 3. This result also contains the *pythagorean inequalities*, well known and celebrated results from information theory.

Section 6 introduces *exponential families* and Theorem 4 establishes their relevance for the easy determination of equilibrium and optimal strategies, in particular *maximum entropy distributions* in cases where the models studied are given by what we call *feasible preparations*.

Section 7, the final section, contains a discussion of various points related to the text with a view also to desirable further research.

Throughout the study, we have emphasized the underlying ideas and though a fair number of proofs are given, we have introduced certain simplifications and tried to avoid mathematical subtleties.

2. Abstract philosophical considerations

The whole is the *world*. We shall consider *situations* from the world which involve *Nature*, without a mind but holder of *truth*, and *Observer*, seeking the truth but restricted to *belief*. In contrast to Nature, Observer has a conscious and creative mind which can be exploited to obtain *knowledge* as effortlessly as possible.

By x we denote a generic *truth instance* associated with a specific situation. We imagine that Nature has "chosen" this instance from a certain set $\mathbb{M}$. To simplify notationally, we shall not express any dependency of $\mathbb{M}$ on the situation. This is justified by the fact that for the present study only one situation will be considered at a time. The set $\mathbb{M}$ is the *set of possible truth instances*.

In any situation, Observer speculates over what Nature is up to, and Observer expresses his belief in the form of an assignment of a *belief instance*, typically denoted by y, to the situation. To simplify, we assume that the belief instance is also chosen from $\mathbb{M}$.

Observers choice of belief instance in any specific situation is considered a subjective choice taking available information into account such as general insight and any specific *prior knowledge*. These thoughts agree with Bayesian thinking, and as such are subject to standard criticism which applies to this line of thought, cf. [3].

Observer acts by designing *experiments* and by making subsequent *observations*. We shall later return in more detail to this aspect. For now we note that the result of observations can be more or less informative, ranging from initial and very limited experience to a final more conclusive stage. It is the final stage we have in mind. We refer to it as *knowledge* and think of it as the *synthesis of extensive experience*.

The end result of Observers endeavours in any particular situation is a *knowledge instance*, typically denoted by z. We reserve the letter $\mathbb{F}$ for the *set of knowledge instances*. We assume that $\mathbb{M} \subseteq \mathbb{F}$. Often, $\mathbb{F} = \mathbb{M}$ will hold.

The connection to extensive experience is just one side of "knowledge". We may also view knowledge as *the way the World presents truth to Observer in any given situation* and, therefore, as the way Observer *perceives* the situation. We assume that knowledge is a function which depends on the situation through the already introduced notions of truth- and belief instances. Formally, we shall operate with a function $\Pi : \mathbb{M} \times \mathbb{M} \to \mathbb{F}$, and interpret $z = \Pi(x, y)$ as the knowledge instance in a situation with x as truth instance and y as belief instance. We call Π the *interactor*, sometimes the *global interactor* in order to distinguish it from a related concept to be introduced later. Two worlds with the same interactor are identified, thus conceived as the *same* world. We use $\mathcal{W}_\Pi$ to denote the world with interactor Π.

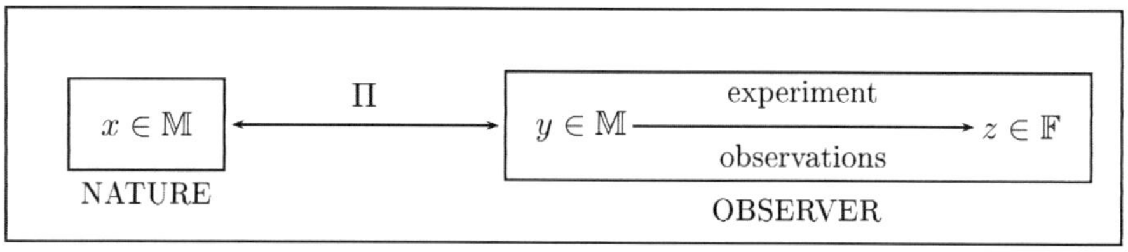

Figure 1. Some key elements of the philosophical considerations

The *classical world*, $\mathcal{W}_1$, is characterized by the interactor Π_1 given by

$$\Pi_1(x, y) = x\,. \tag{1}$$

This is the *world of observable truth* or, expressed somewhat differently, a world where *truth is learnable*. As another extreme, consider $\mathcal{W}_0$, conceived as a *black hole* and characterized by the interactor Π_0 given by

$$\Pi_0(x, y) = y\,. \tag{2}$$

In this world, no matter what Observer does, he will only see a mirror image of himself – it is a world of total narcissism, *what you see is what you believe.* By contrast, in a classical world, *what you see is what is true.*

We also consider mixtures of the two worlds identified above. For this to make sense, we assume that $\mathbb{M}$ is embedded in a linear space. Then, to each $q \in \mathbb{R}$, we may consider the world characterized by the interactor Π_q defined by

$$\Pi_q(x, y) = qx + (1 - q)y\,. \tag{3}$$

The worlds defined in this way are the *Tsallis worlds*, denoted $\mathcal{W}_q$.

All worlds which we will consider will be *sound* in the sense that $\Pi(x, y) = x$ provided there is a *perfect match* (*belief matches truth*), i.e. provided $y = x$.

In order to enable quantitative reasoning, we introduce functions which determine the *effort* needed by Observer in order to acquire knowledge. For reasons discussed in Section 7, these functions are called *score functions.* In principle, any function $\Phi : \mathbb{M} \times \mathbb{M} \to [0, \infty]$ could be a score function [1]. Of course, in order for the intended interpretation to make sense, Φ should be defined in some meaningful way pertaining to the special world and the special situation considered. Only then can $\Phi(x, y)$ be taken to represent the necessary effort needed by Observer in order to gain the knowledge $z = \Pi(x, y)$ in a situation with truth-instance x and belief-instance y. We claim that the appropriate selection of a score function will depend on tools of "description" available to Observer. This view will become more clear when, in the next section, we turn to probabilistic modelling.

We imagine that Observer has many sensible score functions to choose from. Of particular interest are *proper score functions* which are score functions which satisfy the inequality

$$\Phi(x, y) \geq \Phi(x, x) \tag{4}$$

[1] negative values could be allowed and would be convenient for certain wider studies, especially for enabling a smooth treatment of Kullback's *minimum information discrimination principle.*

for any $(x, y) \in \mathbb{M} \times \mathbb{M}$ and for which equality in (4) only holds in case of a perfect match ($y = x$). The implied variational principle, viz. for Observer to choose, whenever possible, a proper score function among the available score functions, we refer to as the *Perfect Match Principle* (PMP). We may allow *singular* cases of PMP for which equality in (4) can take place in other cases than the perfect match case $y = x$. We shall always emphasize if such singular cases are allowed.

The significance of proper score functions can be illuminated by the following considerations for which we change a bit the role of Nature. What we shall imagine is that Nature can communicate with Observer. Then we talk of an *expert* (by the name "Expert") rather than Nature. Consider some situation and assume that Observer seeks the advice of Expert. Experts best advice is x. However, for dubious reasons, Expert may be tempted to give an advice y which differs from x. With the ongoing crisis in world economy in mind, you may think of a bank advisor and a situation involving investment planning. With access to a proper score function, Observer can enter into a contract with Expert which will encourage Expert to be honest, i.e. to give the advice reflecting a perfect match, $y = x$. The contract may be formulated as an agreement for Observer to pay a one-and-for-all sum on signing the contract supplied with a kind of insurance scheme according to which Expert must pay a penalty in the amount of $\Phi(x, y)$ as soon as the true nature of Experts insight, in the form of x, will be revealed to Observer. Clearly, when Φ is a proper score function, it is in Experts own interest to be honest, i.e. to give the advice $y = x$, as this will minimize the penalty to be payed to Observer. Considerations of this nature emerged in statistical studies as detailed in Section 7.

Essential for our analysis are the two concepts, *interaction* (given by Π) and *score* or *effort* (given by Φ). When we work in a world $\mathcal{V}_\Pi$ and Observer has chosen the score function Φ, the resulting model is denoted $\mathcal{V}(\Pi, \Phi)$. Consider such a model and assume that Φ is a proper score function. We may then introduce the important notions *entropy* and *divergence*, here denoted by the letters S and D. Entropy relates to a possible truth instance x, whereas divergence relates to a situation characterized by a pair (x, y) of associated truth- and belief instances:

$$\mathrm{S}(x) = \Phi(x, x) \tag{5}$$

$$\mathrm{D}(x, y) = \Phi(x, y) - \mathrm{S}(x)\,. \tag{6}$$

Thus, conceiving Φ as effort, *entropy is minimal effort* and *divergence is excess* or *redundant effort*. Though infinite values are conceivable, we ignore this problem and simply assume that D can be defined so that $\mathrm{D}(x, y) = 0$ if and only if $y = x$ and so that the *linking identity*

$$\Phi(x, y) = \mathrm{S}(x) + \mathrm{D}(x, y) \tag{7}$$

always holds.

Note that the notions introduced require a proper score function and that entropy is a derived quantity. In consistency with this observation, the thesis that *entropy should never be considered as an isolated quantity* seems to represent a sound and fruitful guiding principle.

The fact that $\mathrm{D}(x, y) \geq 0$ with equality if and only if there is a perfect match is the *fundamental inequality of information theory* (FI), here with information theory understood in a rather general abstract sense.

3. Probabilistic modelling

For our probabilistic modelling, situations are related to a discrete *alphabet* $\mathbb{A}$ and the set of possible truth instances is taken to be the set of probability distributions over $\mathbb{A}$, in symbols $\mathbb{M} = M_+^1(\mathbb{A})$. A typical truth instance is given by the point probabilities: $x = \big(x_i\big)_{i \in \mathbb{A}}$, thus the x_i's are non-negative numbers adding to 1. Similarly, a belief instance is a probability distribution

$y = (y_i)_{i\in\mathbb{A}} \in \mathbb{M}$ [2]. For a knowledge instance $z = (z_i)_{i\in\mathbb{A}}$ we only require that the z_i's are well defined real numbers.

We assume that the interactors to be considered act *locally*, i.e., for some function π defined on $[0,1]\times[0,1]$,

$$\Pi(x,y) = \Big(\pi(x_i.y_i)\Big)_{i\in\mathbb{A}}.$$

The function π is the *local interactor* or just the *interactor*. In order to ensure that the corresponding global interactor is sound, we assume that π is *sound*, i.e. that $\pi(s,s) = s$ for all $0 \le s \le 1$. We also assume that suitable regularity conditions are fulfilled, say continuity and continuous differentiability in the interior domain and, further, that π assumes finite values, except possibly for cases with $\pi(s,t) = \infty$ when $t = 0 < s$.

For the probabilistic models, the world defined by π is denoted $\mathcal{V}_\pi$. The primary example is the *q-interactor* π_q given by

$$\pi_q(s,t) = qs + (1-q)t\,. \tag{8}$$

We put $\mathcal{V}_q = \mathcal{V}_{\pi_q}$. The worlds defined in this way are the *Tsallis worlds* in the probabilistic setting.

The equation (8) expresses a linear (or affine) relationship among probabilities. It is a form of *direct linearity*. It is conceivable that, instead, physical circumstances dictate that a linear relationship only applies to certain function values applied to point probabilities. This points to interactors π_q^ξ of the form

$$\pi_q^\xi(s,t) = \xi^{-1}\Big(\pi_q\big(\xi(s),\xi(t)\big)\Big)\,. \tag{9}$$

For this we require that the function ξ is a smooth strictly increasing function on $[0,1]$. When we work with one of these interactors, we speak about *ξ-linearity* and the associated worlds are denoted $\mathcal{V}_q^\xi$. In mathematical terms, the interactors π_q^ξ are *generalized meanvalues*; for a classical reference, see [4].

Of course, for $q = 1$ or $q = 0$ nothing new is obtained in this way. But for other values of q, interesting interactors emerge. For instance, with $\xi(s) = \ln s$ we find that π_q^ξ determines the *geometric average* $(s,t) \mapsto s^q t^{1-q}$. We write this interactor as π_q^G and the associated world as $\mathcal{V}_q^G$ (G for "geometric"), whereas π_q may be written as π_q^A and $\mathcal{V}_q$ as $\mathcal{V}_q^A$ (A for "arithmetic").

Concerning the value of q, the "cleanest" results are obtained for $0 < q \le 1$, but also values $q > 1$ are of interest. We find negative values of q less interesting for reasons given in Theorem 2.

An interactor π is *weakly consistent* if $\sum_{i\in\mathbb{A}} z_i = 1$ for every pair $(x,y) \in \mathbb{M}\times\mathbb{M}$ with $z = \pi(x,y)$. If even $z \in \mathbb{M}$ can be concluded, the interactor is *strongly consistent.*

Proposition 1. *The only worlds $\mathcal{V}_\pi$ with a weakly consistent interactor π are the Tsallis worlds $\mathcal{V}_q$ for some real value of q, and only for $0 \le q \le 1$ do we obtain a world with a strongly consistent interactor.*

The proof follows a standard pattern and is not given here.

We turn to a closer investigation of worlds $\mathcal{V}_\pi$ with the specification of an associated score function. The first thing to do is to agree on what the sensible score functions for $\mathcal{V}_\pi$ are. Just as for the global interactor, they should act locally. Therefore, the key question is what the "local score" associated with a basic event which has true probability s and believed probability t should be. For this we note that $\pi(s,t)$ is the "force" by which the basic event is perceived

[2] Finer modelling will allow that the set of available belief instances is different from $\mathbb{M}$, or even depends on the actual truth instance; an example of this is to allow that y is an incomplete distribution ($\sum y_i < 1$) and to insist that $y \succ x$ i.e. that $x_i > 0 \Rightarrow y_i > 0$.

by Observer. Then we have to consider what effort Observer should attach to the basic event. As Observer does not know s, this can only depend on Observers own belief, represented by t. Thus, for some function κ, Observer should attach the effort $\kappa(t)$ to a single occurrence of the basic event. This should be multiplied by the perceived force. We conclude that the local contribution to the score function should be $\pi(s,t)\kappa(t)$. Based on such more or less – more, hopefully – convincing intuitive ideas, we agree that every acceptable score function is of the form

$$\Phi(x,y) = \sum_{i \in \mathbb{A}} \pi(x_i, y_i)\kappa(y_i) \tag{10}$$

with κ some function defined on $[0,1]$. The function κ is the *descriptor* and we say that κ *generates* Φ. We may interpret κ as the *cost of information*, viz. $\kappa(t)$ is the price Observer is willing to pay or, in other words, the effort (energy) Observer is willing to allocate in order to obtain the information carried by an event with (believed) probability t. We also refer to $\kappa(t)$ as the *description cost* allocated to a probability-t event, though, for non-classical worlds, we are not able to suggest concrete methods of description – such as those based on *coding* – which could be relevant in this respect. Observe that the score $\Phi(x,y)$ is the *perceived average cost of description* or, as the acquisition of knowledge or *information* is concerned, *the perceived average cost of information*. The corresponding *true average cost*, which is the average allocation of description effort by Observer but seen from the point of view of Nature, is

$$\Phi_{\text{true}}(x,y) = \sum_{i \in \mathbb{A}} x_i \kappa(y_i)\,. \tag{11}$$

On the technical side we impose the following conditions on a descriptor: $\kappa(1) = 0$, κ is continuous on $[0,1]$, finite on $]0,1]$, continuously differentiable on $]0,1]$ and, finally, we impose a condition of *normalization*, viz. that $\kappa'(1) = -1$. The last condition corresponds to a choice of *unit*. The choice made gives *natural units*, nats. Had we, instead, imposed the condition $\kappa'(1) = \ln \frac{1}{2}$, we would have obtained *binary units*, bits. We find that 1 nat ≈ 1.4427 bits.

The descriptor only determines the score function when the interactor is known. Thus, in (10), we may write $\Phi(x,y|\pi,\kappa)$ or similar for clarification. Normally, π and κ will be understood from the context and we may drop these letters from the notation. To specify that we work in the world with interactor π and consider the score function generated by π and the descriptor κ, we refer to this setting as $\mathcal{V}(\pi,\kappa)$.

Consider $\mathcal{V}(\pi,\kappa)$ and assume that the associated score function is proper. Entropy and divergence are then obtained from (5) and (6). We find:

$$\mathrm{S}(x) = \sum_{i \in \mathbb{A}} x_i \kappa(x_i)\,, \tag{12}$$

$$\mathrm{D}(x,y) = \sum_{i \in \mathbb{A}} \Big(\pi(x_i,y_i)\kappa(y_i) - x_i\kappa(x_i)\Big) \tag{13}$$

$$= \sum_{i \in \mathbb{A}} \delta(x_i,y_i)\,, \tag{14}$$

where δ denotes a special function, the *divergence generator* which is defined by

$$\delta(s,t) = \Big(\pi(s,t)\kappa(t) + t\Big) - \Big(s\kappa(s) + s\Big)\,. \tag{15}$$

Clearly, entropy as defined by (12) is a well defined, possibly infinite quantity. The fact that also the two formulas for divergence give well defined quantities is not immediately obvious. In fact this follows in a quite fundamental way, alas depending on a as yet unsolved technical problem. Thus we can only formulate as a conjecture the following statement:

Conjecture 1. *For $\mathcal{V}(\pi, \kappa)$, the score function defined by π and κ is a proper score function if and only if $\delta(s,t) \geq 0$ for every $(s,t) \in [0,1] \times [0,1]$ with equality only for $t = s$.*

The inequality $\delta \geq 0$ (with equality only as stated) we refer to as the *pointwise fundamental inequality* (PFI). Clearly, it implies the fundamental inequality (FI) and thus the properness of the associated score function. In practice, PFI is a much simpler inequality to verify than FI attacked directly. The missing proof of the "only if" part of the conjecture has no practical consequences as all concrete positive results we need are proved via PFI.

It remains to investigate if the descriptor can be adapted to the interactor in a way so that a proper score function emerges. To answer this question, we shall need the function χ defined by

$$\chi(t) = \frac{\partial \pi}{\partial t}(t,t) \tag{16}$$

(so χ is the partial derivative of $\pi(s,t)$ in direction of t, evaluated at the diagonal $s = t$).

Theorem 1. *Let π be an interactor and assume that χ defined by (16) is bounded in the vicinity of 1. Then, there exists at most one descriptor κ which generates a proper score function for the world $\mathcal{V}_\pi$. Indeed, if κ is such a descriptor, κ must be the unique solution to the differential equation*

$$t\kappa'(t) + \chi(t)\kappa(t) + 1 = 0 \tag{17}$$

which satisfies the condition $\kappa(1) = 0$.

Proof. Assume that the score function associated with the descriptor κ is proper. For $0 < t < 1$ put

$$f(t) = t\kappa'(t) + \chi(t)\kappa(t)\,.$$

Consider a probability vector $x = (x_1, x_2, x_3)$ with positive point probabilities. Then the function F given by

$$F(y) = F(y_1, y_2, y_3) = \sum_{i=1}^{3} \pi(x_i, y_i)\kappa(y_i)$$

assumes its minimal value for the interior point $y = x$ when restricted to probability distributions. As standard regularity conditions are fulfilled, there exists a Lagrange multiplier λ such that

$$\frac{\partial}{\partial y_i}\Big(F(y) - \lambda \sum_{i=1}^{3} y_i\Big) = 0 \text{ for } i = 1,2,3$$

when $y = x$. This shows that $f(x_1) = f(x_2) = f(x_3)$.

Using this, first with $(x_1, x_2, x_3) = (\frac{1}{2}, x, \frac{1}{2} - x)$ for a value of x in $]0, \frac{1}{2}[$, and then with $(x_1, x_2, x_3) = (x, \frac{1}{2}(1-x), \frac{1}{2}(1-x))$, one is soon led to the conclusion that f is constant on $]0,1[$. In view of the boundedness condition on χ which is tacitly assumed (weaker conditions will do), the constant must be -1 and the result follows. □

We refer to the one and only possible descriptor which could generate a proper score function for $\mathcal{V}_\pi$ as the *basic candidate* and denote it by $\kappa[\pi]$. As the boundedness condition imposed on χ in Theorem 1 is very weak (perhaps even superfluous), the theorem may be stated briefly by saying that the basic candidate is uniquely defined, given the interactor. The result supports the conjecture. Indeed, if κ generates a proper score function, we may introduce, for each s, the function f_s defined by $f_s(t) = \pi(s,t)\kappa(t) + t$ and then note that by (17), each f_s has a stationary point at $t = s$. As, for every $(x,y) \in \mathbb{M} \times \mathbb{M}$,

$$\sum_{i \in \mathbb{A}} f_{x_i}(y_i) \geq \sum_{i \in \mathbb{A}} f_{x_i}(x_i)\,,$$

it appears plausible that the stationary points of the functions f_s are all minimal points, and this will imply that the conjecture holds.

The conjecture and its relation to PFI tempts us to define an *adjusted* notion of the total description effort, denoted by $\tilde{\Phi}$: $\tilde{\Phi}(x, y) = \sum_{i\in\mathbb{A}}\left(\pi(x_i, y_i)\kappa(y_i) + y_i\right)$. The added terms in the summands, the y_i's, are interpreted as the contributions to the *total overhead* stemming from the respective basic events. Here, "overhead" is related to whatever action, typically observations of an experiment, is involved for Observer in order to acquire the knowledge $z = \Pi(x, y)$. The total overhead in any situation is $\sum y_i = 1$. In other words, the normalization condition $\kappa'(1) = -1$ corresponds to choosing the overhead cost as the unit to work with. Adjusting also the entropy function, one finds that *adjusted entropy* is always bounded below by the overhead cost, 1 nat.

One should of course respect the fact that Theorem 1 only provides a *candidate* for a well behaved descriptor. In each case one has to check if this candidate really provides a descriptor which generates a proper score function. When this is the case, $\kappa = \kappa[\pi]$ is the *ideal descriptor* associated with π and we will typically assume that we work with this descriptor when the world $\mathcal{V}_\pi$ is considered. Notationally we shall write $\mathcal{V}(\pi)$ rather than $\mathcal{V}(\pi, \kappa)$ to indicate this.

In case of the Tsallis worlds, the determination of the basic candidate as well as the necessary checking whether or not it is ideal is straight forward and leads us to a main result. Before stating it, we find it convenient, following Tsallis [5], to introduce the *deformed logarithms* $\ln_q$. For $q \in \mathbb{R}$, they are defined by

$$\ln_q t = \begin{cases} \ln t \text{ if } q = 1 \\ \frac{1}{1-q}\left(t^{1-q} - 1\right) \text{ otherwise}. \end{cases} \tag{18}$$

Theorem 2. *Consider one of the worlds $\mathcal{V}_q$. If $q < 0$, no descriptor defines a proper score function for $\mathcal{V}_q$. If $q \geq 0$, there exists a unique descriptor which defines a proper score function for $\mathcal{V}_q$, viz. the descriptor κ_q given by*

$$\kappa_q(y) = \ln_q \frac{1}{y}. \tag{19}$$

All worlds $\mathcal{V}(\pi_q, \kappa_q)$ with $q > 0$ are regular, whereas the world $\mathcal{V}(\pi_0, \kappa_0)$, a black hole, is singular, in fact divergence vanishes identically in this case.

Proof. The function χ from Theorem 1 is the constant function $1 - q$. Solving (17), you find that $\kappa[\pi_q]$ is given by (19). From simple examples, one finds that for $q < 0$, the perfect match principle does not hold with this descriptor.

It remains to consider values $q \geq 0$. The cases $q = 0$ and $q = 1$ are left to the reader. For the remaining cases, consider the divergence generator

$$\delta_q(s, t) = \frac{q}{1-q} s t^{q-1} + t^q - \frac{1}{1-q} s^q. \tag{20}$$

An application of the geometric-arithmetic mean inequality shows that PFI holds (consider the cases $0 < q < 1$ and $q > 1$ separately and collect the two positive terms). □

Appropriate explicit formulas for score or effort (Φ_q), entropy (S_q), and divergence (D_q) in the Tsallis worlds are easily derived from (10), (12)-(14) and (18)-(20). By Theorem 2, only cases with $q > 0$ yields ideal, non-singular descriptors, hence also proper score functions.

By Theorem 1, you are led in a unique way from an interactor to a descriptor. A natural question arises, that is, if the converse is also true. It is not. Thus, one cannot in general know which world you operate in, i.e. know the interactor, if one only knows the descriptor. Even if the descriptor is ideal, the world cannot be determined. Concrete illustrations of this fact are provided by the following result:

Proposition 2. *For $q > 0$, $\mathcal{V}_q^A$ and $\mathcal{V}_q^G$ both have κ_q given by (19) as ideal descriptor.*

We leave the proof to the reader, only noting that for $q > 1$, the fact that κ_q is the ideal descriptor for $\mathcal{V}_q^G$ is a weaker statement than the similar statement – contained in Theorem 2 – for $\mathcal{V}_q^A$ as then $\pi_q^G \geq \pi_q^A$. For $0 < q < 1$, the statement is stronger as then $\pi_q^G \leq \pi_q^A$. Therefore, the statement requires a proof which is in fact easy to accomplish.

As a consequence of our observation note the following: If we only focus on entropy S_q, and this only needs the descriptor κ_q for its specification, we cannot know the score- or the divergence functions. Both Φ_q^A and D_q^A and Φ_q^G and D_q^G are possible sets of associated functions, the one set corresponding to the interactor π_q^A, the other to π_q^G.

4. Preparations and controls

For this section, as well as for Sections 5 and 6, we work in a world $\mathcal{V} = \mathcal{V}(\pi) = \mathcal{V}(\pi, \kappa)$ with κ the ideal descriptor associated with π. We assume that κ is strictly decreasing on $[0, 1]$. The key example is $\pi = \pi_q$ for a $q > 0$. The situations we have in mind all involve distributions over the same alphabet $\mathbb{A}$.

The main issue we shall now discuss is the philosophical question "what *can* Observer know?". What we mean by this is that in any concrete situation, the set of truth instances chosen by Nature will normally be restricted to some subset of $\mathbb{M}$. Such a subset we call a *preparation.* Typically, we denote a preparation by the letter $\mathcal{P}$. But not every subset of $\mathbb{M}$ can, realisticly, occur. The preparations which can actually be realized, we call the *feasible preparations.* They tell us what Observer can know.

In order to arrive at a meaningful, concrete and operational definition of what a feasible preparation is, we put forward the rough idea that what Observer *can* do to enable the acquisition of knowledge is to fix allocations of effort (or energy) to individual events and also to fix an overall threshold and then enforce Nature to choose only truth instances which respect the indicated constraints.

First, regarding the allocation of efforts, we adopt the view that *belief is a tendency to act.* Therefore, we start from a belief instance ξ and transform it to a more suitable object when having possible actions by Observer in mind. The transformation is denoted $\xi \curvearrowright \hat{\xi}$ with $\hat{\xi}$ defined as the family $\hat{\xi} = \big(\kappa(\xi_i)\big)_{i \in \mathbb{A}}$. We find it convenient to define a *control* as a family $w = (w_i)_{i \in \mathbb{A}}$ for which there exists $\xi \in \mathbb{M}$ such that $w = \hat{\xi}$, i.e. such that $w_i = \kappa(\xi_i)$ for each $i \in \mathbb{A}$.

Secondly, we imagine that, corresponding to a chosen control $w = \hat{\xi}$, Observer can realize a certain *experimental set-up* which consists of various parts such as machinery, instruments and so on, including a special handle which he uses to fix the *level of total effort, h.* The idea then is that this results in a restriction of the truth instances chosen by Nature to the set of $x \in \mathbb{M}$ for which the average effort as perceived by Observer is h, i.e. $\Phi(x, \xi) = h$. The preparation obtained by restricting the truth instances in this way is denoted $\mathcal{P}(w, h)$, i.e. (still with $w = \hat{\xi}$)

$$\mathcal{P}(w, h) = \{x \in \mathbb{M} | \Phi(x, \check{w}) = h\} . \tag{21}$$

Note that these preparations are *level sets* obtained by fixing the belief instance in the score function. Preparations of this form are called the *basic preparations.*

Having chosen a control, the scene is set and observations can begin with the reading of measuring instruments etc. Observer may want to use the same experimental set-up for several experiments by adjusting the level of total effort. Of course, the level should always be set so that $\mathcal{P}(w, h)$ is non-empty.

We can now give a precise definition of a feasible preparation: It is a non-empty finite intersection of basic preparations. The *genus* of a feasible preparation is the smallest number of basic preparations needed to define the preparation. Thus, a *feasible preparation of genus* 1 is the same as a basic preparation.

Regarding the previously introduced concept of "situations", we emphasize that they should always be related to feasible preparations.

By $\mathbb{K}$ we denote the set of controls, i.e. $\mathbb{K} = \{\hat{\xi} | \xi \in \mathbb{M}\}$. The choice of the letter "K" reflects that for the classical world, a control is strongly related to coding. For this reason we may also use the term *coder* instead of control. The transformation $\xi \curvearrowright \hat{\xi}$ is a bijection between $\mathbb{M}$ and $\mathbb{K}$. In order to determine the inverse transformation, denoted $w \curvearrowright \breve{w}$, we need to know the inverse function of κ. This function we call the *probability checker* and denote by $\rho = \rho_\kappa$. We insist that ρ is defined on all of $[0, \infty]$ even though there may be a *cut-off* in κ, as the range $[0, \kappa(0)]$ may be a proper subset of $[0, \infty]$. The formal definition is as follows:

$$\rho(a) = \begin{cases} \kappa^{-1}(a) \text{ if } a \leq \kappa(0) \\ 0 \text{ otherwise.} \end{cases} \tag{22}$$

The probability checker provides a tool to determine how "complex" events you can describe with access to a given number of nats. The lower the probability, the more complex the event. With access to a nats, you can describe any event with a probability as low as $\rho(a)$.

For a Tsallis world with $0 < q \leq 1$, however large your resources to nats, there are events so complex that you cannot describe them, whereas, if $q > 1$, you can describe *any* event if you have access to K nats if only K is large enough ($K \geq \frac{1}{q-1}$).

Another illustration: If, for sample point $i \in \mathbb{A}$ you have decided how many nats you are willing to allocate to i, say a_i nats, this will only be feasible if the *generalized Kraft inequality* holds:

$$\sum_{i \in \mathbb{A}} \rho_\kappa(a_i) \leq 1\,. \tag{23}$$

5. Preparations and games

Consider some preparation $\mathcal{P}$. For the time being, we need not assume that $\mathcal{P}$ is feasible. Observer cannot know which truth instance Nature has chosen, except that it is in $\mathcal{P}$. It lies nearby to speculate if Observer can identify some "typical" element of $\mathcal{P}$, perhaps in order to use that as an appropriate belief instance. The key to actually do so is to view the interplay between Nature and Observer as a *game*. The game should be a two-person zero-sum game with Nature and Observer fighting over description effort. Nature is a maximizer and Observer a minimizer.

The available *strategies* for Nature are limited to truth instances in the preparation $\mathcal{P}$. As for Observer, we have chosen to take the set $\mathbb{K}$ of controls, rather than the set $\mathbb{M}$ of belief instances as strategies. We find this most natural in view of the interpretations. If one wishes, a transformation of definitions and results to involve only belief instances is possible, though at times a bit awkward. In view of our focus on controls, we introduce a variant $\Psi : \mathbb{M} \times \mathbb{K} \to [0, \infty]$ of the score function by agreeing that

$$\Psi(x, w) = \Phi(x, \breve{w})\,. \tag{24}$$

A further consequence of the change of focus concerns divergence which is better conceived as a *redundancy* of effort corresponding to the given control. Denoting redundancy with R, we define

$$\mathrm{R}(x, w) = \mathrm{D}(x, \breve{w})\,. \tag{25}$$

We use the notation $\gamma(\mathcal{P})$ for the game defined above. The two *values* of the game are defined as usual, cf. [6] or [7], for instance. The value seen from the point of view of Nature is

$$\sup_{x \in \mathcal{P}} \inf_{w \in \mathbb{K}} \Psi(x, w)\,. \tag{26}$$

As the infimum here is nothing but the entropy $\mathrm{S}(x)$, we find that the value given by (26) is the *maximum entropy value* associated with $\mathcal{P}$:

$$\mathrm{S}_{\max}(\mathcal{P}) = \sup_{x\in\mathcal{P}} \mathrm{S}(x)\,. \tag{27}$$

As for the other value of the game, it is given by

$$\inf_{w\in\mathbb{K}} \sup_{x\in\mathcal{P}} \Psi(x,w)\,. \tag{28}$$

For every $w \in \mathbb{K}$, the associated *individual risk* is defined by

$$\mathrm{Ri}(w) = \mathrm{Ri}(w|\mathcal{P}) = \sup_{x\in\mathcal{P}} \Psi(x,w)\,. \tag{29}$$

The *minimal risk value* associated with $\mathcal{P}$ is then defined as

$$\mathrm{Ri}_{\min}(\mathcal{P}) = \inf_{w\in\mathbb{K}} \mathrm{Ri}(w)\,, \tag{30}$$

i.e. as the value given by (28).

Clearly, $\mathrm{S}_{\max}(\mathcal{P}) \leq \mathrm{Ri}_{\min}(\mathcal{P})$. This is the *minimax inequality*. If equality holds and defines a finite quantity, the game is in *game theoretical equilibrium* or just in *equilibrium*. An *optimal strategy for Nature* is the same as a truth instance in $\mathcal{P}$ with maximal entropy. An *optimal strategy for Observer* is the same as a control w with $\mathrm{Ri}(w) = \mathrm{Ri}_{\min}$.

Another concept of equilibrium is related to *robustness*: A control w^* is *robust* if, for some $h \in \mathbb{R}$, $\Psi(x,w^*) = h$ for all $x \in \mathcal{P}$. The number h is the *level of robustness*. If Observer chooses a robust strategy, the description effort will be independent of which strategy Nature has chosen. There is an important connection between the two concepts of equilibria:

Theorem 3 (robustness and pythagorean inequalities). *Assume that $x^* \in \mathcal{P}$ and that $w^* = \hat{x^*}$ is robust with robustness level h. Then $\gamma(\mathcal{P})$ is in equilibrium with h as value. Furthermore, for any $x \in \mathcal{P}$,*

$$\mathrm{S}(x) + \mathrm{R}(x,w^*) \leq \mathrm{S}_{\max}(\mathcal{P}) \tag{31}$$

and, for every control w,

$$\mathrm{Ri}(w) \geq \mathrm{S}_{\max}(\mathcal{P}) + \mathrm{R}(x^*,w)\,. \tag{32}$$

Proof. Though the argument is known and easy, cf. Theorem 6.2 of [8], we present it here too. Firstly, by assumption, $\mathrm{Ri}(w^*) = h = \Psi(x^*,w^*) = \mathrm{S}(x^*)$, hence the game is in equilibrium with x^* as optimal strategy for Nature and w^* as optimal strategy for Observer.

For any $x \in \mathcal{P}$ we find that $\mathrm{S}(x) \leq \mathrm{S}(x) + \mathrm{R}(x,w^*) = \Psi(x,w^*) = h$, i.e. (31) holds. And for any control w, $\mathrm{Ri}(w) \geq \Psi(x^*,w)$, hence (32) follows from the linking identity (7). □

Equation (31), in the form $\mathrm{S}(x) + \mathrm{D}(x,x^*) \leq \mathrm{S}_{\max}(\mathcal{P})$, is the *pythagorean inequality* going back to Čencov [9] and Csiszár [10], and (32) is the *reverse pythagorean inequality*. From a game theoretical point of view, these inequalities are trivial consequences of Nash's *saddle value inequalities*.

6. Exponential families

We shall develop a simple criterion which facilitates the identification of situations of equilibrium and also facilitates the search for optimal strategies for both Nature and Observer. The first result will help us understand why the level sets play a central role. Actually, we first run into *sub-level sets*, defined by

$$\mathcal{P}_{\leq}(w,h) = \{x|\Psi(x,w) \leq h\} \tag{33}$$

Proposition 3. *Consider a truth instance x^* and a control w^*. Then a necessary and sufficient condition that there exists a preparation $\mathcal{P}$ for which the game $\gamma(\mathcal{P})$ is in equilibrium and has x^* and w^* as optimal strategies, is that $h^* = \Psi(x^*,w^*) < \infty$ and that $w^* = \hat{x^*}$. And when these conditions are fulfilled, the largest such set is the sub-level set $\mathcal{P}_{\leq}(w^*,h^*)$.*

Proof. (Compare with [8], Theorem 8.2). If, for some $\mathcal{P}$, $\gamma(\mathcal{P})$ is in equilibrium with x^* and w^* as optimal strategies then, by *Nash's saddle value inequalities,*

$$\Psi(x,w^*) \leq \Psi(x^*,w^*) \leq \Psi(x^*,w) \tag{34}$$

for every $x \in \mathcal{P}$ and every control w. Furthermore, $\Psi(x^*,w^*)$ is finite. Let $y^* = \check{w^*}$. Then, by the right hand inequality of (34), and as Φ is a proper score function, $\Phi(x^*,x^*) \geq \Phi(x^*,y^*) \geq \Phi(x^*,x^*)$, hence $\Phi(x^*,y^*) = \Phi(x^*,x^*)$ and we conclude that $y^* = x^*$. It follows that $w^* = \hat{x^*}$. It only remains to remark that the left hand inequality of (34) implies directly that $\mathcal{P} \subseteq \mathcal{P}_{\leq}(w^*,h^*)$ and the first part of the proof is complete.

For the second half, assume that $h^* = \Psi(x^*,w^*) < \infty$ and that $w^* = \hat{x^*}$. From the latter condition, the saddle value inequalities (34) are verified for the preparation $\mathcal{P}_{\leq}(w^*,h^*)$ and then, using also the former condition, equilibrium follows as well as optimality of the strategies x^* and w^*. □

The result points to a unique role for the sub-level sets. However, we hold the view that it is the highest level which is of relevance, and hence we stick to (intersections of) level sets, rather than sub-level sets. Also, this will allow us to exploit results on robustness.

These considerations lead to an expedient approach to the most common type of problems within statistical physics as well as many other branches of science where questions of entropy optimization and equilibrium come up. First, for a finite set $\mathbf{w} = (w_1,\cdots,w_n)$ of controls and a corresponding set of levels, $\mathbf{h} = (h_1,\cdots,h_n)$, we put $\mathcal{P}(\mathbf{w},\mathbf{h}) = \cap_{\nu=1}^{n}\mathcal{P}(w_\nu,h_\nu)$ (if non-empty, this is a feasible preparation of genus at most n) and for $\mathbf{w}$ fixed, we denote by $\mathbb{P}(\mathbf{w})$ the family of all feasible preparations of this form. The corresponding *exponential family*, denoted by $\hat{\mathcal{E}}(\mathbf{w})$, is the set of controls ε which are robust for all preparations in the family $\mathbb{P}(\mathbf{w})$. In terms of belief instances, the exponential family is the family $\mathcal{E}(\mathbf{w})$ of all belief instances ξ for which $\xi = \check{\varepsilon}$ for some $\varepsilon \in \mathcal{E}(\mathbf{w})$.

From the definitions introduced and from Theorem 3 on robustness you find the following simple, but useful result:

Theorem 4. *Consider a family $\mathbf{w} = (w_1,\cdots,w_n)$ of controls and the associated family $\mathbb{P}(\mathbf{w})$ of preparations. Let x^* be a truth instance, ε^* the corresponding control ($\varepsilon^* = \hat{x^*}$) and assume that $\varepsilon^* \in \hat{\mathcal{E}}(\mathbf{w})$. Put $\mathbf{h} = (h_1,\cdots,h_n)$ with $h_\nu = \Psi(x^*,w_\nu)$ for $\nu = 1,\cdots,n$. Then $\gamma(\mathcal{P}(\mathbf{w},\mathbf{h}))$ is in equilibrium and has x^* and w^* as optimal strategies. In particular, x^* is the maximum entropy distribution for the preparation $\mathcal{P}(\mathbf{w},\mathbf{h})$.*

A simplification occurs in case the world we consider is one of the Tsallis worlds $\mathcal{V}_q = \mathcal{V}(\pi_q,\kappa_q)$ for a $q > 0$ as then one can identify controls in the exponential family. The reason is that in view of the linear character of π_q, we realize that every control ε of the form $\alpha + \sum_1^n \beta_\nu w_\nu$ for

Journal of Physics: Conference Series **201** (2010) 012013 doi:10.1088/1742-6596/201/1/012013

suitable constants α and $\beta_1, \cdots, \beta_n$ is a member of $\check{\mathcal{E}}(\mathbf{w})$. Typically, for given $\beta_1, \cdots, \beta_n$ you adapt α to these numbers so that $\alpha + \sum_1^n \beta_\nu w_\nu$ is a genuine control. This requires that

$$\sum_{i \in \mathbb{A}} \rho_q\big(\alpha + \beta_1 w_{1,i} + \cdots + \beta_n w_{n,i}\big) = 1\,, \tag{35}$$

where ρ_q is the probability checker for κ_q, cf. (23). Note that for the classical case $q = 1$, an extra simplification is that, given the β's, α can be solved from this equation.

7. Discussion

The present study has been preceeded by [11] and [12], and will, according to plan, be followed by a more detailed publication. However, it may well be that the present exposition is the ideal starting point for any reader who will embark on research in the direction taken. In particular, we have made certain simplifying assumptions and also, we have toned down the very abstract considerations and mainly stayed within more standard probabilistic modelling.

The main result on the identification of Tsallis entropy, Theorem 2, is derived based on the assumption that, firstly, one should allow for an interaction between truth, belief and knowledge and, secondly, one should accept a rather innocent variational principle, that optimal performance is obtained when there is a perfect match between truth and belief. It should be emphasized that though these principles could, by some adversary, be viewed as axioms, they are intended as key elements of an operational interpretation. Further studies may justify this view, possibly based on a deeper insight into physics than here displayed. It is a bit disturbing to the author that, according to previous studies (also reported in the recent monograph [2]), Tsallis entropy with q negative is of importance for the discussion of certain physical phenomena, whereas our approach does not really accept negative values of q.

It is noted that our study is devoid of a dynamical dimension. The only implicit dynamical element is the natural succession in time of truth, belief and knowledge (via experiments and observations). Again, further research is needed on this point.

Another direction of research, which should fit well into the pronounced philosophical basis, would be an extension to also cover the quantum setting.

The author has put great emphasis on the choice of terminology. However, especially concerning one issue, the proper naming of what here appears as score functions, there may well be other attractive options. The term *inaccuracy*, following Kerridge [13], did not appeal to the author, *effort function* hits better the intended interpretation and is kept "in reserve" whereas *complexity* is overloaded and *description function* or similar is fine for the classical setting, but with no clarification of an eventual role of more physical description (or coding) outside the classical case, it would appear a bit premature. Finally, *score function* and *proper score function* were chosen as these notions appear in statistical decision theory and has a long history which can be traced from Csiszár, [14]. There, and in works referred to, one will also find what we have termed the Perfect Match Principle. One disadvantage with the term "score" is that it gives the impression of something attractive seen with Observers eyes, quite opposite to what is the case. Of course, one could say that a high score is attractive when seen with the eyes of Nature if you follow the game theoretical line of thought. Or one could say that when Observer *has* acquired knowledge, the score can be taken as a quantitative measure of the information gained. The situation is similar to the one concerning the appearance of both entropy and "negentropy" in certain texts. Without citing extensively from the literature on score functions, we point to [15] by Good, mainly because there we found the view that *belief is a tendency to act*, a view which fits nicely into the philosophy here adopted. As a final comment, note that when restricted to the classical setting, the consideration of proper scoring functions gives an elegant introduction to Shannon entropy which is more descriptive than axiomatic, we find, and which provides an attractive supplement to the more common approaches via coding.

As already indicated, further research on the fundamental nature of the quantities characterized is much desired. This also concerns a more complete interpretation of descriptors outside the classical case. In this connection, [16], [17] and references there as well as [18] may be relevant.

Regarding the, admittedly, rather brief treatment of maximum entropy optimization in Section 6, note that this does not rely on the introduction of Lagrange multipliers, but uses a more direct approach in line with [19]. Also note that the usual introduction of *partition functions* does not, apparently, generalize in an adequate way when we study generalized entropies. In contrast, the *log-partition function* does generalize in a natural way, viz. as the solution α to (35).

Further concrete results on proper scoring functions may be developed by going outside the probabilistic setting, say by considering more general real valued functions than distributions and other concrete generalized meanvalues than the two chosen here, the arithmetic- and the geometric meanvalues.

The connection (duality) between feasible preparations and exponential families might fit well into geometric ideas as developed by Amari and his school, cf. [20]. Other possibilities to apply duality considerations are rather obvious, but not developed here. In this connection, a somewhat different and purely axiomatic approach developed in a preliminary setting in [21] should also be of relevance.

The untraditional approach in Section 6 to the much studied notion of exponential families is emphasized. It appears to be natural with reference to reasonably sound interpretations (via controls etc.) and also, the approach applies in other contexts than probabilistic ones, e.g. related to geometry (an application to Sylvesters problem of location theory is briefly indicated in [22]). The key role of game theory is evident. To the author, it is surprising that general and rather simple results as developed by Nash are so powerful – beyond what many researchers have presently realized, it seems. For the classical case, the game theoretical approach was studied in [8]. From that reference, we point to the discussion after Corollary 4.2 and to the important Theorem 6.9 which shows that no opportunities are missed when focusing on robustness as is done in our definition of exponential families.

For other works which emphasize the role of exponential families for statistical physics, see Naudts, [23] and [24]. These works have some similarity with the present approach.

Finally, some comments on the appearance of Tsallis entropy. The first publications are Havrda and Charvát [25] and, independently, Daróczy [26]. The latter author emphasized a characterization via functional equations, cf. also [27] and the more recent reference work [28]. The first appearance in the physical literature is due to Lindhard and Nielsen [29]. Subsequently, Lindhard gave a careful treatment of aspects of the measuring process, cf. [30], which may also be of relevance to our treatment. The trend-setting publication [1] from 1988 by Tsallis marks the efficient introduction of the generalized entropies within the physical community. At the time of publication, Tsallis was unaware of the earlier research. Regarding [29] and [30], these papers were largely ignored, though there is a casual reference to Lindhard's work in [31].

The success of Tsallis in launching the entropy measures which now bear his name is due to the direct approach and the fact that when combined with Jaynes *Maximum Entropy Principle*, cf. [32], main problems of statistical physics lead to *power laws*, a popular class of distributions when heavy-tailed distributions are needed.

The present approach is in line with earlier game theoretical considerations, cf. [19]. Because of a relation to Bregman divergences, we also point the reader to [23] and works referred to there.

Acknowledgments
Bjarne Andresen introduced me to Tsallis entropy, Imre Csiszár provided hints to the literature on score functions and Stig Stenstrup pointed me to the works by Lindhard and Nielsen. Interest and discussions with Jan Naudts, Hiroki Suyari and Constantino Tsallis before and during the workshop in Kyoto on generalized entropies is gratefully acknowledged, as are comments, mainly on terminology and philosophy, from Peter Harremoës and Nisheeth Srivastava.

References

[1] Tsallis, C. Possible generalization of Boltzmann-Gibbs statistics. *J. Stat. Physics* **1988**, *52*, 479–487.
[2] Tsallis, C. *Introduction to Nonextensive Statistical Mechanics.* Springer, Berlin Heidelberg, 2009.
[3] Wikipedia contributers. Bayesian probability **2009**.
[4] Hardy, G. H.; Littlewood, J. E.; Pólya, G. *Inequalities.* Cambridge University Press, second edition Ed., 1959.
[5] Tsallis, C. What are the numbers that experiments provide? *Quimica Nova* **1994**, *17*, 468.
[6] Aubin, J. P. *Optima and equilibria. An introduction to nonlinear analysis.* Springer, Berlin, 1993.
[7] Cesa-Bianchi, N.; Lugosi, G. *Prediction, learning and games.* Cambridge University Press, Cambridge, 2006.
[8] Harremoës, P.; Topsøe, F. Maximum entropy fundamentals. *Entropy*(Sept. 2001), *3*, 191–226.
[9] Čencov, N. N. *Statistical Decision Rules and Optimal Inference.* Nauka, Moscow, 1972. In russian, translation in "Translations of Mathematical Monographs", 53. AmericanMathematical Society, 1982.
[10] Csiszár, I. I-divergence geometry of probability distributions and minimization problems. *Ann. Probab.* **1975**, *3*, 146–158.
[11] Topsøe, F. Truth, Belief and Experience – a route to Information. *Journal of Contemporary Mathematical Analysis – Armen. Acad. Scienc.* **2009**, *44*, 105–110.
[12] Topsøe, F. On truth, belief and knowledge. In *2009 IEEE International Symposium on Information Theory.* IEEE, Washington, (June 2009), pp. 139–143.
[13] Kerridge, D. F. Inaccuracy and inference. *J. Roy. Stat. Soc. B.* **1961**, *23*, 184–194.
[14] Csiszár, I. Axiomatic Characterizations of Information Measures. *Entropy* **2008**, *10*, 261–273.
[15] Good, I. J. Rationel decisions. *J. Royal Statist. Soc., Series B* **1952**, *14*, 107–114.
[16] van der Lubbe, J. C. A. On certain coding theorems for the information of order α and oftypeβ. In *Trans. Eighth Prague Conf. on Inform. Theory, Statist. Decision Functions,RandomProcesses.* Czech. Acad. Science, Academia Publ., Prague, 1978. Prague, 1979.
[17] Ahlswede, R. Identification Entropy, In *General Theory of Information Transfer and Combinatorics*; *Lecture Notes in Computer Science*, *4123*, 595–613. Springer, Berlin, 2006.
[18] Suyari, H. Tsallis entropy as a lower bound of average description length for the q-generalized code tree. In *Proceedings ISIT 2007.* IEEE, 2007, pp. 901–905.
[19] Topsøe, F. Exponential Families and MaxEnt Calculations for Entropy Measures of Statistical Physics. In *Complexity, Metastability, and Non-Extensivity, CTNEXT07*; Abe, Hermann, Q. R. T., Ed., 2007, Vol. 965, *AIP Conference Proceedings*, pp. 104–113.
[20] Amari, S.; Nagaoka, H. *Methods of Information Geometry*, Vol. 191, *Translations of Mathematical monographs.* Oxford University Press, 2000.
[21] Topsøe, F. Game Theoretical Optimization inspired by Information Theory. *J. Global Optim.* **2009**, pp. 553–564.
[22] Topsøe, F. Intrinsic Methods for Optimization Problems. In *2008 IEEE International Symposium on Information Theory.* IEEE, Washington, (July 2008), pp. 1627–1630.
[23] Naudts, J. Generalised exponential families and associated entropy functions. *Entropy* **2008**, *10*, 131–149.
[24] Naudts, J. The q-exponential family in statistical physics. *Cent. Eur. J. Phys.* **2009**, *7*, 405–413.
[25] Havrda, J.; Charvát, F. Quantification method of classification processes. Concept of structural a-entropy. *Kybernetika* **1967**, *3*, 30–35.
[26] Daróczy, Z. Generalized Information Functions. *Information and Control* **1970**, *16*, 36–51.
[27] Aczél, J.; Daróczy, Z. *On measures of information and their characterizations.* Academic Press, New York, 1975.
[28] B. Ebanks, P. S.; Sander, W. *Characterizations of Information Measures.* World Scientific, Singapore, 1998.
[29] Lindhard, J.; Nielsen, V. Studies in Statistical Dynamics. *Mat. Fys. Medd. Dan. Vid. Selsk.* **1971**, *38*, 1–42.
[30] Lindhard, J. On the Theory of Measurement and its Consequences in Statistical Dynamics. *Mat. Fys. Medd. Dan. Vid. Selsk.* **1974**, *39*, 1–39.
[31] Jaynes, E. T. Where do we Stand on Maximum Entropy?, In *The Maximum Entrropy Formalism*; Levine, R.; Tribus, M., Eds., pp. 1–104. M.I.T. Press, Cambridge, MA, 1979.
[32] Jaynes, E. T. Information theory and statistical mechanics, I and II. *Physical Reviews* **1957**, *106 and 108*, 620–630 and 171–190.

Journal of Physics: Conference Series **201** (2010) 012014 doi:10.1088/1742-6596/201/1/012014

Characterization for entropy and solution of set function on lattice

Aoi Honda, Yoshiaki Okazaki

Kyushu Institute of Technology, 680-4 Kawazu, Iizuka City, Fukuoka 820-8502, JAPAN

E-mail: aoi@ces.kyutech.ac.jp

E-mail: okazaki@ces.kyutech.ac.jp

Abstract. We propose a definition for the entropy of a monotone set function defined on a lattice which are not necessarily the whole power set, but satisfy the condition of regularity. Our definition encompasses the classical definition of Shannon for probability measures, as well as the definition of Marichal for classical fuzzy measures and may have applicability to most fuzzy measures which appear in applications. We give also an axiomatization of this entropy. This axiomatization is in the spirit of Faddeev's axiomatization for the classical Shannon entropy. After that, using same idea we introduce a generalization of the Shapley value for a set function defined on a lattice and give two types of necessary and sufficient conditions.

1. Introduction

The classical definition of the Shannon entropy [11] for probability measures is at the core of information theory. Therefore, many attempts for defining an entropy for a set function more general than the classical probability measures have been done, in particular for the monotone set function, what is called the fuzzy measure [13] or the capacity. A first attempt by Yager [15] and Marichal and Roubens [10] proposed a definition having suitable properties, and which can be considered as the generalization of the Shannon entropy. Another attempt was also done by Dukhovny [2], in a different spirit. All these works considered a finite universal set, and the power set as underlying set system. Whereas we consider more general fuzzy measure, defined on lattices or set systems satisfying a kind of regularity, i.e., that all maximal chains have the same length, which are not necessary the whole power set [5]. Our definition encompasses the classical definition of Shannon, as well as the definition of Marichal [10], and may have applicability to most monotone set functions which appear in applications. Moreover our characterization of this entropy is in the spirit of Faddeev's axiomatization for the classical Shannon entropy [6]. In this paper, we introduce a definition for the entropy of the fuzzy measure defined on lattices, which is a yet more general case of monotone set functions. A set system is merely a collection of subsets of some universal set, containing the empty set and the universal set itself. The difference with usual fuzzy measures on power set is that the fuzzy measure, with which we are concerned, is not defined for every subset (or coalition, in a game theoretic perspective). We define the entropy for such fuzzy measures on set systems, provided that the set system satisfies a regularity condition. Moreover, using join-irreducible elements, there is a close connection between set systems and lattices, so that our definition can be applied to monotone set function on lattices as well. We give also a characterization for this generalized entropy.

Using the same idea, we can define a solution of game defined on a lattice. The second aim of the paper is to provide a definition of a solution of game on lattice, which encompasses the Shapley value. We give two types of axiomatic characterizations of a solution for game on a lattice.

2. Preliminaries

Throughout this paper, we consider a finite universal set $N = \{1, 2, \ldots, n\}$, $n \geq 1$, and 2^N denotes the power set of N.

Definition 1 (set system) *Let $\mathfrak{N}$ be a subset of 2^N. If $\mathfrak{N}$ contains $\emptyset$ and N, then we call $(N, \mathfrak{N})$ (or simply $\mathfrak{N}$ if no confusion occurs) a* set system.

Let $A, B \in \mathfrak{N}$. We say that A is *covered* by B, and write $A \prec B$ or $B \succ A$, if $A \subsetneq B$ and $A \subseteq C \subsetneq B$ together with $C \in \mathfrak{N}$ imply $C = A$.

Definition 2 (maximal chain) *Let $\mathfrak{N}$ be a set system. We call $\mathfrak{C}$ a* maximal chain of $\mathfrak{N}$ *if $\mathfrak{C} = (C_0, C_1, \ldots, C_m)$ satisfies $\emptyset = C_0 \prec C_1 \prec \cdots \prec C_m = N, C_i \in \mathfrak{N}, i = 0, \ldots, m$.*

The length of the maximal chain $\mathfrak{C} = (C_0, C_1, \ldots, C_m)$ is m. We denote the set of all maximal chains of $\mathfrak{N}$ by $\mathcal{M}(\mathfrak{N})$.

Definition 3 (totally ordered set system, chain) *We say that $(N, \mathfrak{N})$ is a* totally ordered set system *or a* chain *if for any $A, B \in \mathfrak{N}$, either $A \subseteq B$ or $A \supsetneq B$.*

If $(N, \mathfrak{N})$ is a totally ordered set system, then it has only one maximal chain.

Definition 4 (regular set system) *We say that $(N, \mathfrak{N})$ is a* regular set system *if for any $A, B \in \mathfrak{N}$ satisfying $A \succ B$, $|A \setminus B| = 1$ holds.*

Definition 5 (fuzzy measure on a set system) *Let $(N, \mathfrak{N})$ be a set system. A function $v : \mathfrak{N} \to [0, 1]$ is a* fuzzy measure on $(N, \mathfrak{N})$ *if it satisfies $v(\emptyset) = 0, v(N) = 1$ and for any $A, B \in \mathfrak{N}, v(A) \leq v(B)$ whenever $A \subseteq B$.*

For clarifying the domain of v, we denote often the triplet $(N, \mathfrak{N}, v)$ instead of simply v. we denotes it by the triplet $(N, \mathfrak{N}, v)$. $\mathcal{F}(N, \mathfrak{N})$ denotes the set of all fuzzy measure defined on $(N, \mathfrak{N})$.

For $v \in \mathcal{F}(N, \mathfrak{N})$ and $\mathfrak{C} := (C_0, C_1, \ldots, C_m) \in \mathcal{M}(\mathfrak{N})$, define $p^{v,\mathfrak{C}}$ by

$$\begin{aligned} p^{v,\mathfrak{C}} &:= (p_1^{v,\mathfrak{C}}, p_2^{v,\mathfrak{C}}, \ldots, p_m^{v,\mathfrak{C}}) \\ &= (v(C_1) - v(C_0), v(C_2) - v(C_1), \ldots, v(C_m) - v(C_{m-1})). \end{aligned}$$

Note that $p^{v,\mathfrak{C}}$ satisfies $p_i^{v,\mathfrak{C}} \geq 0, i = 1, \ldots, m$ and $\sum_{i=1}^m p_i^{v,\mathfrak{C}} = 1$.

We turn now to definitions of the entropy. We first recall the classical definition of Shannon.

Definition 6 (Shannon entropy[11]) *For a probability measure $p = (p_1, \ldots, p_n)$ on N, the* Shannon entropy *of p is defined by*

$$H_S(p) = H_S(p_1, \ldots, p_n) := -\sum_{i=1}^n p_i \log p_i,$$

where $p_i := p(\{i\})$ and $\log$ denoting the base 2 logarithm, and by convention $0 \log 0 := 0$.

Marichal has generalized the Shannon entropy for applying to the fuzzy measure on 2^N.

Definition 7 (Marichal's entropy [10]) *For $v \in \mathcal{F}(N, 2^N)$, Marichal's entropy of v is defined by*

$$H_M(v) = H_M(N, 2^N, v) := - \sum_{A \subseteq N \setminus \{i\}} \gamma^n_{|A|}(v(A \cup \{i\}) - v(A)) \log (v(A \cup \{i\}) - v(A)),$$

where

$$\gamma^n_k := \frac{(n-k-1)!k!}{n!}.$$

The axiomatization of Marichal's entropy is proposed by Kojadinovic et al. [9]. Their axiomatization is however rather complicated, due to the presence of a recursive axiom, whose meaning is hard to grasp.

Using the concept of the maximal chain, we have proposed a generalization of the definition of the entropy for fuzzy measures defined on regular set systems.

Definition 8 (entropy of fuzzy measure on set system [5]) *Let $(N, \mathfrak{N})$ be a regular set system. For $v \in \mathcal{F}(N, \mathfrak{N})$, the entropy of v is defined by*

$$H_{HG}(v) = H_{HG}(N, \mathfrak{N}, v) := \frac{1}{|\mathcal{M}(\mathfrak{N})|} \sum_{\mathfrak{C} \in \mathcal{M}(\mathfrak{N})} H_S(p^{v,\mathfrak{C}}).$$

We discuss the domains of H. Let Σ_n be a all regular set of $N = \{1, 2, \ldots, n\}$. The domain of $\mathcal{F}_{\mathcal{R}} := \bigcup_{n=1}^{\infty} \bigcup_{\mathfrak{N} \in \Sigma_n} \mathcal{F}(N, \mathfrak{N})$, that is, all of the fuzzy measures defined on regular set systems.

We introduce further concepts about fuzzy measures, which will be useful for stating axioms of the entropy.

Definition 9 (dual measure) *For $v \in \mathcal{F}_{\mathcal{R}}$, the* dual measure *of v is defined on $\mathfrak{N}^d := \{A \in 2^N \mid A^c \in \mathfrak{N}\}$ by $v^d(A) := 1 - v(A^c)$ for any $A \in \mathfrak{N}^d$, where $A^c := N \setminus A$.*

Definition 10 (permutation of v) *Let π be a permutation on N. For $v \in \mathcal{F}_{\mathcal{R}}$, the* permutation of v *by π is defined on $\pi(\mathfrak{N}) := \{\pi(A) \in 2^N \mid A \in \mathfrak{N}\}$ by $\pi \circ v(A) := v(\pi^{-1}(A))$.*

Let us consider a chain of length 2 as a set system, denoted by $\mathbf{2}$ (e.g., $\{\emptyset, \{1\}, \{1, 2\}\}$), and a fuzzy measure $v^{\mathbf{2}}$ on it. We denote by the triplet $(0, u, 1)$ the values of $v^{\mathbf{2}}$ along the chain and we suppose $\mathbf{2} := \{\emptyset, \{1\}, \{1, 2\}\}$ unless otherwise noted.

Definition 11 (embedding of $v^{\mathbf{2}}$) *Let $(N, \mathfrak{C})$ be a totally ordered regular set system such that $\mathfrak{C} := \{C_0, \ldots, C_n\}$, $C_{i-1} \prec C_i$, $i = 1, \ldots, n$. For $v \in \mathcal{F}(N, \mathfrak{C})$, $v^{\mathbf{2}} = (0, u, 1)$ and $C_k \in \mathfrak{C}$, v^{C_k} is called the* embedding *of $v^{\mathbf{2}}$ into v at C_k and defined on the totally ordered regular set system $(N^{C_k}, \mathfrak{C}^{C_k})$ by*

$$v^{C_k}(A) := \begin{cases} v(A), & \text{if } A = C_j, j < k, \\ v(C_{k-1}) + u \cdot \big(v(C_k) - v(C_{k-1})\big), & \text{if } A = C'_k \\ v(C_{j-1}), & \text{if } A = C'_j, j > k, \end{cases}$$

where $\{i_k\} := C_k \setminus C_{k-1}, i'_k \neq i''_k$, $(N \setminus \{i_k\}) \cap \{i'_k, i''_k\} = \emptyset, N^{C_k} := (N \setminus \{i_k\}) \cup \{i'_k, i''_k\}, C'_k := (C_k \setminus \{i_k\}) \cup \{i'_k\}, C'_j := (C_{j-1} \setminus \{i_k\}) \cup \{i'_k, i''_k\}$ for $j > k$, and $\mathfrak{C}^{C_k} := \{C_0, \ldots, C_{k-1}, C'_k, C'_{k+1}, \ldots, C'_{n+1}\}$.

Journal of Physics: Conference Series **201** (2010) 012014 doi:10.1088/1742-6596/201/1/012014

Definition 12 (join-irreducible element) *An element $x \in (L, \leq)$ is* join-irreducible *if for all $a, b \in L$, $x \neq \bot$ and $x = a \vee b$ implies $x = a$ or $x = b$.*

$\mathcal{J}(L)$ denotes the set of all join-irreducible elements of L. The mapping η for any $a \in L$, defined by

$$\eta(a) := \{x \in \mathcal{J}(L) \mid x \leq a\}$$

is a lattice-isomorphism of L onto $\eta(L) := \{\eta(a) \mid a \in L\}$, that is, $(L, \leq) \cong (\eta(L), \subseteq)$. In this paper, though we consider only set functions defined on set systems, using the translation η, we can treat set functions defined on lattice $(L, \leq)$ as one defined on set system $(\mathcal{J}(L), \eta(L))$.

3. Axiomatization of the entropy of fuzzy measures

We introduce five axioms for the entropy of fuzzy measures.

Axiom 1 (continuity) *The function $f(u) := H(0, u, 1) = H(v^{\mathbf{2}}, \{1, 2\}, \mathbf{2})$ is continuous on $[0, 1]$, and there exists $u_0 \in [0, 1]$ such that $f(u_0) > 0$.*

Axiom 2 (dual invariance) *For any $v^{\mathbf{2}} = (0, u, 1) \in \mathcal{F}(\{1, 2\}, \mathbf{2})$,*

$$H(0, u, 1) = H(0, 1 - u, 1).$$

Axiom 3 (increase by embedding) *Let $(N, \mathfrak{C})$ be a totally ordered regular set system. For any $v \in \mathcal{F}(N, \mathcal{C})$, any $C_k \in \mathfrak{C}$ and any $v^{\mathbf{2}} = (0, u, 1)$, the entropy of v^{C_k} is*

$$H(v^{C_k}) = H(v) + (v(C_k) - v(C_{k-1})) \cdot H(0, u, 1).$$

Axiom 4 (convexity) *Let $(N, \mathfrak{N})$, $(N, \mathfrak{N}_1), \ldots, (N, \mathfrak{N}_k)$ be regular set systems satisfying $\mathcal{M}(\mathfrak{N}) = \bigcup_{j=1}^{k} \mathcal{M}(\mathfrak{N}_j)$, and $\mathcal{M}(\mathfrak{N}_i) \cap \mathcal{M}(\mathfrak{N}_j) = \emptyset$ for all $i \neq j$. Then there exists $(\alpha_1, \ldots, \alpha_k) \in]0, 1[^k$ with $\sum_{j=1}^{k} \alpha_j = 1$ such that for any $v \in \mathcal{F}(N, \mathfrak{N})$, it holds that*

$$H(v) = \alpha_1 H(v|_{\mathfrak{N}_1}) + \cdots + \alpha_k H(v|_{\mathfrak{N}_k}).$$

Axiom 5 (permutation invariance) *For any $v \in \mathcal{F}_{\mathcal{R}}$ and any permutation π on N satisfying $\pi(\mathfrak{N}) = \mathfrak{N}$, it holds that $H(v) = H(\pi \circ v)$.*

Theorem 13 ([6]) *Let $(N, \mathfrak{N})$ be a regular set system and $(N, \mathfrak{N}, v)$ a fuzzy measure. Then there exists the unique function satisfying* Axioms 1, 2, 3, 4 *and* 5, *and it is given by H_{HG}.*

Now we recall the Tsallis entropy. The Tsallis entropy was introduced in 1988 ([14]), and it has been studied intensively by many authors.

Definition 14 *Let q be a positive real number. For a probability measure $p = (p_1, \ldots, p_n)$ on N, the Tsallis entropy of p is defined by*

$$H_T(p) = H_T(p_1, \ldots, p_n) = -\sum_{i=1}^{n} p_i^q \ln_q p_i,$$

where $p_i := p(\{i\})$ and

$$\ln_q(x) := \frac{x^{1-q} - 1}{1 - q}.$$

We can define the Tsallis entropy of the fuzzy measure defined on set systems, using our framework.

Definition 15 *Let $(N, \mathfrak{N})$ be a regular set system. For $v \in \mathcal{F}(N, \mathfrak{N})$ the entropy of v is defined by*

$$H_T(v) = H_T(N, \mathfrak{N}, v) := \frac{\lambda_q}{|\mathcal{M}(\mathfrak{N})|} \sum_{\mathfrak{C} \in \mathcal{M}_n(\mathfrak{N})} H_T(p^{v,\mathfrak{C}}),$$

where λ_p is a positive constant that depends on q.

Modifying Axiom 3 as follows, we obtain an axiomatization of ($\mathrm{H_T}$) (see also [3]).

Axiom 3' *Let $(N, \mathfrak{C})$ be a totally ordered regular set system. For any $v \in \mathcal{F}(N, \mathfrak{C})$, any $C_k \in \mathfrak{C}$ and any $v^{\mathbf{2}} = (0, u, 1)$, the entropy of v^{C_k} is*

$$H(v^{C_k}) = H(v) + \big(v(C_k) - v(C_{k-1})\big)^q \cdot H(0, u, 1).$$

Theorem 16 *Let $(N, \mathfrak{N})$ be a regular set system and $(N, \mathfrak{N}, v)$ a fuzzy measure. Then there exists the unique function satisfying* Axioms 1, 2, 3', 4 *and* 5, *and it is given by H_T.*

4. Solution of cooperative game

We call a function $v : \mathfrak{N} \to I\!R$ a *characteristic function* of the cooperative game if it satisfies $v(\emptyset) = 0$. Now we consider N is a set consisting of n players. Then a subset of N is called a coalition, and the characteristic function v means the profit by coalitions of n players. We denote the characteristic function by the triplet $(N, \mathfrak{N}, v)$ in the same manner as the fuzzy measure and we call it *generalized cooperative game* or simply *game*. $(N, 2^N, v)$ is a classical cooperative game. The solution is the function from the whole set of the game v to n-dimension real which measures each player's contribution or share-out. The Shapley value is the most important concept as a solution of the game, and they are characterized by natural axiomatizations [12]. We have generalized the Shapley value for applying games defined on $\mathfrak{N} \subseteq 2^N$ which satisfies a kind of regularity, not only defined on 2^N [5][7]. Our generalization make the Shapley value applicable to more general games, for example multi-choice game [8], bi-capacity [4], games on anti-matroid [1]. We have also shown an axiom system which consists of six axioms characterizing this generalized the Shapley value.

We denote the set of all games defined on the regular set system $(N, \mathfrak{N})$ by $\mathcal{G}(N, \mathfrak{N})$. Let Σ_n be a all regular set of $N = \{1, 2, \ldots, n\}$ and define $\mathcal{G}_\mathcal{R} := \bigcup_{n=1}^{\infty} \bigcup_{\mathfrak{N} \in \Sigma_n} \mathcal{G}(N, \mathfrak{N})$, that is, all of the games defined on regular set systems, and $\mathcal{G}_\mathcal{P} := \bigcup_{n=1}^{\infty} \mathcal{G}(N, 2^N)$, that is, all of the games defined on power sets.

Definition 17 (original Shapley value [12]) *For $v \in \mathcal{G}_\mathcal{P}$, the* Shapley value *of v, $\Phi_{\mathrm{sh}}(v) = \Phi_{\mathrm{sh}}(N, 2^N, v) := (\phi^1_{\mathrm{sh}}(v), \ldots, \phi^n_{\mathrm{sh}}(v)) \in [0, 1]^n$ is defined by*

$$\phi^i_{\mathrm{sh}}(v) = \phi^i_{\mathrm{sh}}(N, 2^N, v) := \sum_{E \subseteq N \setminus \{i\}} \gamma^n_{|E|}(v(E \cup \{i\}) - v(E)),\ i = 1, \ldots, n.$$

We denote $\Phi(N, \mathfrak{N}, v)$ and $\phi^i(N, \mathfrak{N}, v)$ instead of $\Phi(v)$ and $\phi^i(v)$ for clarifying the domain of v.

It is known that the Shapley value is represented with maximal chains as follows.

$$\phi^i_{\mathrm{sh}}(v, N, 2^N) = \frac{1}{|\mathcal{M}(2^N)|} \sum_{\mathfrak{C} \in \mathcal{M}(2^N)} (v(\mathfrak{C}_*(i) \cup \{i\}) - v(\mathfrak{C}_*(i))), \quad i = 1, \ldots, n,$$

where $\mathfrak{C}_*(i) := \sup\{C \in \mathfrak{C} \mid C \not\ni i\}$.

Faigle and Kern had defined a value for applying the generalized cooperative game using the concept of the maximal chain. We extend their value for applying to more general cases.

Definition 18 (Generalization of Shapley value) *For* $v \in \mathcal{G}_{\mathcal{R}}$*, the solution* $\Psi(v) = \Psi(N, \mathfrak{N}, v) := (\psi^1(v), \ldots, \psi^n(v)) \in I\!R^n$ *of* $(N, \mathfrak{N}, v)$ *is defined by*

$$\psi^i(v, N, \mathfrak{N}) := \frac{1}{|\mathcal{M}_n(\mathfrak{N})|} \sum_{\mathfrak{C} \in \mathcal{M}_n(\mathfrak{N})} (v(\mathfrak{C}_*(i) \cup \{i\}) - v(\mathfrak{C}_*(i))), \quad i = 1, \ldots, n.$$

The original Shapley value is characterized by several reasonable axioms.

Axiom 6 (efficiency) *For any* $v \in \mathcal{G}_{\mathcal{P}}$,

$$\sum_{i \in N} \phi^i(v) = v(N).$$

Axiom 7 (0-valued for null player) *Fix* $v \in \mathcal{G}_{\mathcal{P}}$*. For any null player* $i \in N$*, i.e.,* $v(A \cup \{i\}) = v(A)$ *for any* $A \in N \setminus \{i\}$*,* $\phi^i(v) = 0$ *holds.*

Axiom 8 (symmetry) *If* $i, j \in N$ *are symmetry of* $v \in \mathcal{G}_{\mathcal{P}}$*, i.e.,* $v(A \cup \{i\}) = v(A \cup \{i\})$ *for any* $A \in N \setminus \{i, j\}$*,* $\phi^i(v) = \phi^j(v)$ *holds.*

Axiom 9 (linearity) *For any* $v, w \in \mathcal{G}_{\mathcal{P}}$*,* $\Phi(v + w) = \Phi(v) + \Phi(w)$ *holds.*

Theorem 19 *Let* $(N, 2^N, v)$ *be a game. Then there exists the unique function* $\Phi : \mathcal{G}_{\mathcal{P}} \to I\!R^n$ *satisfying* Axioms 6, 7, 8 *and* 9*, and it is given by* Φ_{sh}.

For yielding to our generalized value, we generalize axioms the above and add one more axiom. We denote the set of all game defined on n-length chain $(N, \mathfrak{C})$ by $\mathcal{G}(N, \mathfrak{C})$ and define $\mathcal{G}_C := \bigcup_{n=1}^{\infty} \mathcal{G}(N, \mathfrak{C})$, that is, all of the games defined on regular set systems which are chains.

Axiom 6' (efficiency) *For any* $v \in \mathcal{G}_{\mathcal{P}} \cup \mathcal{G}_C$

$$\sum_{i \in N} \phi^i(v) = v(N).$$

Definition 20 (null-player of v on the set system) *i is called* null-player *if* $i \in N$ *satisfies that* $v(A \cup \{i\}) = v(A)$ *whenever* $A \in \mathfrak{N}, A \cup \{i\} \in \mathfrak{N}$*,* $\phi^i(v) = 0$.

Axiom 7' (0-valued for null-player) *Fix* $v \in \mathcal{G}_{\mathcal{P}} \cup \mathcal{G}_C$*. For any null-player* $i \in N$*,* $\phi^i(v) = 0$ *holds.*

Axiom 8' (symmetry) *Let* σ *be a permutation on* N*. For any* $v \in \mathcal{G}_{\mathcal{P}} \cup \mathcal{G}_C$,

$$\phi^i(N, 2^N, v) = \phi^{\sigma(i)}(N, 2^N, \sigma \circ v), \quad \phi^i(N, \mathfrak{C}, v) = \phi^{\sigma(i)}(N, \sigma(\mathfrak{C}), \sigma \circ v),$$

where $\sigma(\mathfrak{C}) := \{\sigma(A) \mid A \in \mathfrak{C}\}, \sigma \circ v(S) := v(\sigma^{-1}(S)), S \in \sigma(\mathfrak{C})$.

Axiom 9' (linearity) *For any* $v, w \in \mathcal{G}_{\mathcal{P}} \cup \mathcal{G}_C$*,* $\Phi(v + w) = \Phi(v) + \Phi(w)$ *holds.*

For yielding to our generalized value, we generalize axioms the above and add one more axiom.

Axiom 10 (convexity) Let $(N, \mathfrak{N})$, $(N, \mathfrak{N}_1), (N, \mathfrak{N}_2), \ldots, (N, \mathfrak{N}_m)$ be regular set systems satisfying $\mathcal{M}(\mathfrak{N}) = \mathcal{M}(\mathfrak{N}_1) \cup \cdots \cup \mathcal{M}(\mathfrak{N}_m)$ and $\mathcal{M}(\mathfrak{N}_1) \cap \cdots \cap \mathcal{M}(\mathfrak{N}_m) = \emptyset$ and v be a game on $\mathfrak{N}$. Then there exists $\alpha_1, \ldots, \alpha_m \in]0, 1[$, with $\sum_{k=1}^{m} \alpha_k = 1$ such that for every game $v \in \mathcal{G}(N, \mathfrak{N})$,

$$\Phi(v) = \alpha_1 \Phi(v|_{\mathfrak{N}_1}) + \cdots + \alpha_m \Phi(v|_{\mathfrak{N}_m}).$$

Theorem 21 *Let $(N, \mathfrak{N})$ be a regular set system and $(N, \mathfrak{N}, v)$ a game. Then there exists the unique function satisfying* Axioms 6', 7', 8', 9' *and* 10, *and it is given by* Ψ.

Our new axioms 6', 7', 8' and 9' include axioms 6, 7, 8 and 9, respectively, and assertion of the Theorem 21 includes the theorem 19, so that our new system of axioms is suitable for a solution of the cooperative game.

We treat cooperative games defined on regular set systems. Most cooperative games which appear in applications can be regarded as games on regular set systems using by a kind of translations (See [5]).

Proof of Theorem 21 Assume that Φ satisfies axioms 6',7', 8', 9' and 10.
(i) Case $\mathfrak{N}$ is a chain. Define $v_j \in \mathcal{G}_{\mathcal{C}}$ as

$$v_j(S) := \begin{cases} 1, & |S| \geq j, \\ 0, & \text{otherwise.} \end{cases}$$

By axioms 6' and 7', for any $\lambda \in I\!R$, it holds that

$$\phi^i(\lambda v_j) = \begin{cases} \lambda, & \{i\} = C_j \setminus C_{j-1}, \\ 0, & \text{otherwise,} \end{cases}$$

that is,

$$\phi^i(\lambda v_j) = \lambda v_j(\mathfrak{C}_*(i) \cup \{i\}) - \lambda v_j(\mathfrak{C}_*(i)).$$

$v \in \mathcal{G}_{\mathcal{C}}$ is identified by $v(C_1), \ldots, v(C_n)$, so that we can regard $v \in \mathcal{G}_{\mathcal{C}}$ as the element of n-dimension vector space. Since the linear independency of $v_j, j = 1, \ldots, n$, for $v \in \mathcal{G}_{\mathcal{C}}$, there exist $\lambda_j, j = 1, \ldots n$ with

$$v = \sum_{j=1}^{n} \lambda_j v_j$$

such that

$$v(C_j) = \sum_{j=i}^{n} \lambda j.$$

By axiom 9', for any $v \in \mathcal{G}_{\mathcal{C}}$, it holds that

$$\begin{aligned} \phi^i(v) &= \phi^i\left(\sum_{j=1}^{n} \lambda_j v_j\right) = \sum_{j=1}^{n} \phi^i(\lambda_j v_j) = \sum_{j=1}^{n} \{\lambda_j v_j(\mathfrak{C}_*(i) \cup \{i\}) - \lambda_j v_j(\mathfrak{C}_*(i))\} \\ &= \sum_{j=i}^{n} \{\lambda_j v_j(\mathfrak{C}_*(i) \cup \{i\}) - \lambda_j v_j(\mathfrak{C}_*(i))\} = v(\mathfrak{C}_*(i) \cup \{i\}) - v(\mathfrak{C}_*(i)) \end{aligned}$$

and $\Phi = \Psi$.
(ii) Case $\mathfrak{N} = 2^N$. By axiom 10, there exist $\alpha_{\mathfrak{C}} \in [0, 1], \mathfrak{C} \in \mathcal{M}(2^N)$ with $\sum_{\mathfrak{C} \in \mathcal{M}(2^N)} \alpha_{\mathfrak{C}} = 1$ such that for any $v \in \mathcal{G}_{\mathcal{P}}$

$$\phi^i(v) = \sum_{\mathfrak{C} \in \mathcal{M}(2^N)} \alpha_{\mathfrak{C}} \phi^i(v|_{\mathfrak{C}}) \tag{1}$$

holds.

Define $\mathfrak{N}_{\{i\}} := \{S \in 2^N \mid S \subseteq \{i\} \text{ or } S \supsetneq \{i\}\}, i = 1, \ldots, n$, then by axiom 10 there exist $\beta_{1,1}, \ldots, \beta_{1,n}$ with $\beta_{1,1} + \cdots + \beta_{1,n} = 1$ such that for any $v \in \mathcal{G}_{\mathcal{P}}$,

$$\phi^i(v) = \beta_{1,1}\phi^i(v|_{\mathfrak{N}_{\{1\}}}) + \cdots + \beta_{1,n}\phi^i(v|_{\mathfrak{N}_{\{n\}}})$$

Journal of Physics: Conference Series **201** (2010) 012014 doi:10.1088/1742-6596/201/1/012014

holds.

Assume that $\beta_{1,1}, \ldots, \beta_{1,n}$ are not unique and there exists another representation

$$\phi^i(v) = \beta'_{1,1}\phi^i(v|_{\mathfrak{N}_{\{1\}}}) + \cdots + \beta'_{1,n}\phi^i(v|_{\mathfrak{N}_{\{n\}}}).$$

Defining $v_j \in \mathcal{G}_{\mathcal{P}}$ as

$$v_j(S) := \begin{cases} 1, & |S| \geq j, \\ 0, & \text{otherwise}, \end{cases}$$

players except i are null-players of $v_1|_{\mathfrak{N}_{\{i\}}}$, so that for any $j \neq i$, $\phi^i(v|_{\mathfrak{N}_{\{j\}}}) = 0$ holds and

$$\phi^i(v_1) = \beta_{1,i}\phi^1(v_1|_{\mathfrak{N}_{\{i\}}}), \quad \phi^i(v_1) = \beta'_{1,i}\phi^1(v_1|_{\mathfrak{N}_{\{i\}}}).$$

Hence $\beta_{1,i} = \beta'_{1,i}$, so that $\beta_{1,i}$ is unique. Moreover let σ be a permutation of players i_1 and i_2, then by axiom 8' for any $i_1, i_2 \in N$,

$$\begin{aligned} \phi^{i_1}(N, \mathfrak{N}_{\{i_1\}}, v_1|_{\{i_1\}}) &= \phi^{\sigma(i_1)}(N, \sigma(\mathfrak{N}_{\{i_1\}}), \sigma \circ v_1|_{\{i_1\}}) \\ &= \phi^{i_2}(N, \mathfrak{N}_{\{i_2\}}, v_1|_{\{i_2\}}) \end{aligned}$$

and since any players i_1 and $i_2 \in N$ are symmetry of v_1, for any $i = 1, \ldots, n$,

$$\phi^i(v_1) = \beta_{1,i}\phi^1(v_1|_{\mathfrak{N}_{\{1\}}}) = \phi^1(v_1),$$

so that we obtain $\beta_{1,1} = \cdots = \beta_{1,n} = 1/n$. Next define $\mathfrak{N}_{\{1,i\}} := \{S \in \mathfrak{N}_{\{1\}} \mid S \subseteq \{1,i\} \text{ or } S \supsetneq \{1,i\}\}, i = 2, \ldots, n$, then by axiom 10 there exist $\beta_{2,2}, \ldots, \beta_{2,n}$ with $\beta_{2,2} + \cdots + \beta_{2,n} = 1$, such that for any $v|_{\mathfrak{N}_{\{1\}}} \in \mathcal{G}(N, \mathfrak{N}_{\{1\}})$, it holds that

$$\phi^i(v|_{\mathfrak{N}_{\{1\}}}) \;=\; \beta_{2,2}\phi^i(v|_{\mathfrak{N}_{\{1,2\}}}) + \cdots + \beta_{2,n}\phi^i(v|_{\mathfrak{N}_{\{1,n\}}}).$$

Players except i are null-players of $v_2|_{\mathfrak{N}_{\{2,i\}}}$, so that for any $j \neq i$, $\phi^i(v|_{\mathfrak{N}_{\{1,j\}}}) = 0$ holds and since for $i = 2, \ldots, n$, we have $\phi^i(v_2|_{\mathfrak{N}_{\{1\}}}) = \beta_{2,i}\phi^i(v_2|_{\mathfrak{N}_{\{1,i\}}})$, $\beta_{2,i}$ is unique, and by axiom 8' we have $\beta_{2,2} = \cdots = \beta_{2,n} = 1/(n-1)$. Similarly, define $\mathfrak{N}_{\{1,\ldots,k,i\}} := \{S \in \mathfrak{N}_{\{1,\ldots,k\}} \mid S \subseteq \{1, \ldots, k, i\} \text{ or } S \supsetneq \{1, \ldots, k, i\}\}$, then we have generally for $i = k+1, \ldots, n$, by axiom 10 there exist $\beta_{k+1,k+1}, \ldots, \beta_{k+1,n}$ with $\beta_{k+1,k+1} + \cdots + \beta_{k+1,n} = 1$ such that for any $v|_{\mathfrak{N}_{\{1,\ldots,k\}}} \in \mathcal{G}(N, \mathfrak{N}_{\{1,\ldots,k\}})$,

$$\phi^i(v|_{\mathfrak{N}_{\{1,\ldots,k\}}}) = \beta_{k+1,k+1}\phi^i(v|_{\mathfrak{N}_{\{1,\ldots,k,k+1\}}}) + \cdots + \beta_{k+1,n}\phi^i(v|_{\mathfrak{N}_{\{1,\ldots,k,n\}}})$$

holds.

Players except i are null-players of $v_{k+1}|_{\{1,\ldots,k,i\}}$, so that for any $j \neq i$, $\phi^i(v|_{\mathfrak{N}_{\{1,\ldots,k,j\}}}) = 0$ holds and since for $i = k+1, \ldots, n$, we have

$$\begin{aligned} \phi^i(v_{k+1}|_{\mathfrak{N}_{\{1,\ldots,k\}}}) &= \beta_{k+1,k+1}\phi^i(v_{k+1}|_{\mathfrak{N}_{\{1,\ldots,k,k+1\}}}) + \cdots + \beta_{k+1,n}\phi^i(v_{k+1}|_{\mathfrak{N}_{\{1,\ldots,k,n\}}}) \\ &= \beta_{k+1,i}\phi^i(v_{k+1}|_{\mathfrak{N}_{\{1,\ldots,k,i\}}}), \end{aligned}$$

$\beta_{k+1,i}$ is unique, and by axiom 8' we have $\beta_{k+1,k+1} = \cdots = \beta_{k+1,n} = 1/(n-k)$.

For other terms, decomposing the set system to chains, we obtain

$$\begin{aligned}
\phi^i(v) &= \beta_{1,1}\phi^i(v|_{\mathfrak{N}_{\{1\}}}) + \cdots + \beta_{1,n}\phi^i(v|_{\mathfrak{N}_{\{n\}}}) \\
&= \beta_{1,1}\left[\beta_{2,2}\phi^i(v|_{\mathfrak{N}_{\{1,2\}}}) + \cdots + \beta_{2,n}\phi^i(v|_{\mathfrak{N}_{\{1,n\}}})\right] + \beta_{1,2}\phi^i(v|_{\mathfrak{N}_{\{2\}}}) + \cdots + \beta_{1,n}\phi^i(v|_{\mathfrak{N}_{\{n\}}}) \\
&\ \ \vdots \\
&= \beta_{1,1}\cdots\beta_{n-1,n-1}\phi^i(v|_{\mathfrak{N}_{\{1,\dots,n-1\}}}) + \beta_{1,1}\cdots\beta_{n-1,n-2}\beta_{n-1,n}\phi^i(v|_{\mathfrak{N}_{\{1,\dots,n-2,n\}}}) \\
&\qquad + \sum_{\mathfrak{C}\in\mathcal{M}(2^N)\setminus(\mathfrak{C}_1\cup\mathfrak{C}_2)} \beta\phi^i(v|_{\mathfrak{C}}) \\
&= \frac{1}{n!}\phi^i(v|_{\mathfrak{C}_1}) + \frac{1}{n!}\phi^i(v|_{\mathfrak{C}_2}) + \sum_{\mathfrak{C}\in\mathcal{M}(2^N)\setminus(\mathfrak{C}_1\cup\mathfrak{C}_2)} \beta\phi^i(v|_{\mathfrak{C}}),
\end{aligned}$$

where $\mathfrak{C}_1 := \{\emptyset, \{1\}, \{1,2\}, \dots, N\}, \mathfrak{C}_2 = \{\emptyset, \{1\}, \{1,2\}, \dots, \{1,\dots,n-2\}, \{1,\dots,n-2,n\}, N\}$.

Therefore for (1) $\alpha_{\mathfrak{C}_1} = \alpha_{\mathfrak{C}_2} = 1/n!$ are obtained. Similary $\alpha_{\mathfrak{C}}, \mathfrak{C} \in \mathcal{M}(2^N) \setminus \{\mathfrak{C}_1, \mathfrak{C}_2\}$ is obtained uniquely as $1/n!$.

Consequently, it holds that

$$\phi^i(v) = \frac{1}{|\mathcal{M}(2^N)|}\phi^i(v|_{\mathfrak{C}}), \tag{2}$$

and $\Phi = \Psi$.

(iii) Case $\mathfrak{N}$ is a general set system. Let $\mathfrak{N} \subsetneq 2^N$. Since

$$\mathcal{M}(2^N) = \mathcal{M}(\mathfrak{N}) \cup \bigcup_{\mathfrak{C}\in\mathcal{M}(2^N)\setminus\mathcal{M}(\mathfrak{N})} \mathfrak{C},$$

by axiom 10 there exist $\alpha, \beta_{\mathfrak{C}}, \mathfrak{C} \in \mathcal{M}(2^N) \setminus \mathcal{M}(\mathfrak{N})$ with

$$\alpha + \sum_{\mathfrak{C}\in\mathcal{M}(2^N)\setminus\mathcal{M}(\mathfrak{N})} \beta_{\mathfrak{C}} = 1 \tag{3}$$

such that for any $v \in \mathcal{G}_{\mathcal{P}}$

$$\begin{aligned}
\phi^i(v) &= \alpha\phi^i(v|_{\mathfrak{N}}) + \sum_{\substack{\mathfrak{C}\in \\ \mathcal{M}(2^N)\setminus\mathcal{M}(\mathfrak{N})}} \beta_{\mathfrak{C}}\phi^i(v|_{\mathfrak{C}}) \qquad (4) \\
&= \alpha\left(\sum_{\mathfrak{C}\in\mathcal{M}(\mathfrak{N})} \alpha_{\mathfrak{C}}\phi^i(v|_{\mathfrak{C}})\right) + \sum_{\mathfrak{C}\in\mathcal{M}(2^N)\setminus\mathcal{M}(\mathfrak{N})} \beta_{\mathfrak{C}}\phi^i(v|_{\mathfrak{C}}) = \sum_{\mathfrak{C}\in\mathcal{M}(2^N)} \beta_{\mathfrak{C}}\phi^i(v|_{\mathfrak{C}}),
\end{aligned}$$

where for $\mathfrak{C} \in \mathcal{M}(\mathfrak{N})$ $\beta_{\mathfrak{C}} := \alpha\alpha_{\mathfrak{C}}$, holds. By (2) for $\mathfrak{C} \in \mathcal{M}(2^N) \setminus \mathcal{M}(\mathfrak{N})$ $\beta_{\mathfrak{C}} = 1/n!$ and for $\mathfrak{C} \in \mathcal{M}(\mathfrak{N})$ $\beta_{\mathfrak{C}} = \alpha\alpha_{\mathfrak{C}} = 1/n!$, and since (3)

$$\alpha = 1 - \frac{1}{n!}\cdot|\mathcal{M}(2^N) \setminus \mathcal{M}(\mathfrak{N})| = 1 - \frac{n! - |\mathcal{M}(\mathfrak{N})|}{n!} = \frac{|\mathcal{M}(\mathfrak{N})|}{n!}, \tag{5}$$

substituting (5) for (4), we obtain

$$\phi^i(v|_{\mathfrak{N}}) = \frac{1}{\alpha}\left(\phi^i(v) - \sum_{\mathfrak{C}\in\mathcal{M}(2^N)\setminus\mathcal{M}(\mathfrak{N})} \frac{1}{n!}\phi^i(v|_{\mathfrak{C}})\right) = \frac{1}{|\mathcal{M}(\mathfrak{N})|}\sum_{\mathfrak{C}\in\mathcal{M}(\mathfrak{N})} \phi^i(v|_{\mathfrak{C}}).$$

□

On the other hand, we can give another different type of axiomatization for the generalization of Shapley value, which can be characterized with the same concept of the entropy.

Axiom 11 (continuity) The function the function $f(s,t) := \phi_1(0,s,t)$ is continuous for s on $\mathbb{R}$.

Axiom 12 (efficiency) For any game $(0,s,t)$ on **2**, $\phi_1(0,s,t) + \phi_2(0,s,t) = v(N) = t$.

Axiom 13 (dual invariance) For any $(0,s,t)$, $\Phi(0,s,t) = \Phi(0,s,t)^d$ holds.

Axiom 14 (embedding efficiency) For any $v \in \mathcal{G}_C$, any $(0,s,1)$ and any $C_k \in \mathfrak{C}$, $\phi_i(v^{C_k}) = \phi_i(v)$ for any $i \neq i'_k, i''_k$, $\phi_{i'_k}(v^{C_k}) = \phi_{i_k}(v) \cdot \phi_1(0,s,1)$ and $\phi_{i''_k}(v^{C_k}) = \phi_{i_k}(v) \cdot \phi_2(0,s,1)$ hold, where $\{i_k\} := C_k \setminus C_{k-1} = \{i'_k, i''_k\}$.

Axiom 15 (permutation invariance) For any $v \in \mathcal{G}_{\mathcal{R}}$ and any permutation π on N satisfying $\pi(\mathfrak{N}) = \mathfrak{N}$, $\phi_i(v) = \phi_{\pi(i)}(\pi \circ v), i = 1, \ldots, n$ holds.

We obtain the following theorem:

Theorem 22 ([7]) *Let $(N, \mathfrak{N})$ be a regular set system and $(N, \mathfrak{N}, v)$ a game. Then there exists the unique function satisfying* Axioms 11, 12, 13, 14 *and* 15, *and it is given by Ψ.*

[1] E. Algaba, J.M. Bilbao, R. van den Brink and A. Jime' nez-Losada, Axiomatizations of the Shapley value for cooperative games on antimatroids, Math. Meth. Oper. Res., Vol. 57, 49–65, 2003.

[2] A. Dukhovny, General entropy of general measures, *Internat. J. Uncertain. Fuzziness Knowledge-Based Systems*, **10** (2002), 213–225.

[3] S. Furuichi, On uniqueness theorems for Tsallis entropy and Tsallis relative entropy, *IEEE Trans. Inf. Theory*, **51**, no. 10 (2005), 3638–3645.

[4] M. Grabisch and Ch. Labreuche, Bi-capacities for decision making on bipolar scales, In EUROFUSE Workshop on Information Systems, Varenna, Italy, September 2002.

[5] A. Honda, M. Grabisch, Entropy of capacities on lattices, *Information Sciences*, **176** (2006), 3472-3489.

[6] A. Honda, M. Grabisch, An axiomatization of entropy of capacities on set systems, *European Journal of Operational Research*, **190** (2008), 526-538.

[7] A. Honda, Y. Okazaki, Axiomatization of Shapley value of Faige and Kern type on set systems, Journal of Advanced Computational Intelligence and Intelligent Informatics, Vol. 12 No. 5, 2008, 409–415, 2008.

[8] C.-R. Hsiao and T.E.S. Raghavan, Shapley value for multi-choice cooperative games, I, Games and Economic Behaviour, 5, 240–256, 1993.

[9] I. Kojadinovic, J.-L. Marichal, M. Roubens, An axiomatic approach to the definition of the entropy of a discrete Choquet capacity, *Information Sciences*, 172 (2005), 131-153.

[10] J.-L. Marichal, M. Roubens, Entropy of discrete fuzzy measure, *Internat. J. Uncertain. Fuzziness Knowledge-Based Systems*, **8** (2000), 625–640.

[11] C.E. Shannon, A mathematical theory of communication, *Bell System Tech. Journ.*, **27** (1948), 374–423, 623–656.

[12] L.S. Shapley, A value for n-person games, Kuhn HW, Tucker, AW (eds) Contributions to the Theory of Games Vol. II, Princeton, 307-317, 1953.

[13] M. Sugeno, Fuzzy measures and fuzzy integrals: a survey, in M.M. Gupta, G.N. Saridis, and B.R. Gains, editors, Fuzzy automata and decision processes, pp. 89-102, North Holland, Amsterdam, 1977.

[14] C. Tsallis, Possible generalization of Boltzmann-Gibbs statistics, *J. Stat. Phys.*, **52** (1988), 479–487.

[15] R.R. Yager, On the entropy of fuzzy measures, *IEEE Transaction on Fuzzy Systems*, **8** (2000), 453–461.

Journal of Physics: Conference Series **201** (2010) 012015 doi:10.1088/1742-6596/201/1/012015

Wigner-Yanase-Dyson Skew Information And Uncertainty Relation

Kenjiro Yanagi
Division of Applied Mathematical Science, Graduate School of Science and Engineering, Yamaguchi University, Ube, 755-8611 Japan
E-mail: yanagi@yamaguchi-u.ac.jp

Abstract. We give a trace inequality related to the uncertainty relation of Wigner-Yanase-Dyson skew information. This inequality corresponds to a generalization of the uncertainty relation derived by S.Luo [8] for the quantum uncertainty quantity excluding the classical mixture. Finally we show a counter-example of the uncertainty relation related to Wigner-Yanase-Dyson skew information given in [6] recently.

1. Introduction

Wigner-Yanase skew information

$$\begin{aligned} I_\rho(H) &= \frac{1}{2}Tr\left[\left(i\left[\rho^{1/2}, H\right]\right)^2\right] \\ &= Tr[\rho H^2] - Tr[\rho^{1/2}H\rho^{1/2}H] \end{aligned}$$

was defined in [10]. This quantity can be considered as a kind of the degree for non-commutativity between a quantum state ρ and an observable H. Here we denote the commutator by $[X, Y] = XY - YX$. This quantity was generalized by Dyson

$$\begin{aligned} I_{\rho,\alpha}(H) &= \frac{1}{2}Tr[(i[\rho^\alpha, H])(i[\rho^{1-\alpha}, H])] \\ &= Tr[\rho H^2] - Tr[\rho^\alpha H\rho^{1-\alpha}H], \alpha \in [0, 1] \end{aligned}$$

which is known as the Wigner-Yanase-Dyson skew information. It is famous that the convexity of $I_{\rho,\alpha}(H)$ with respect to ρ was successfully proven by E.H.Lieb in [7]. From the physical point of view, an observable H is generally considered to be an unbounded opetrator, however in the present paper, unless otherwise stated, we consider $H \in B(\mathcal{H})$ represents the set of all bounded linear operators on the Hilbert space $\mathcal{H}$, as a mathematical interest. We also denote the set of all self-adjoint operators (observables) by $\mathcal{L}_h(\mathcal{H})$ and the set of all density operators (quantum states) by $\mathcal{S}(\mathcal{H})$ on the Hilbert space $\mathcal{H}$. The relation between the Wigner-Yanase skew information and the uncertainty relation was studied in [9]. Moreover the relation between the Wigner-Yanase-Dyson skew information and the uncertainty relation was studied in [5, 11]. In our paper [11], we defined a generalized skew information and then derived a kind of an uncertainty relation. In the section 2, we discuss various properties of the Wigner-Yanase-Dyson skew information. Finally in section 3, we give our main result and its proof.

2. Uncertainty relations of Wigner-Yanase-Dyson skew information

We review the relation between the Wigner-Yanase skew information and the uncertainty relation. In quantum mechanical system, the expectation value of an observable H in a quantum state ρ is expressed by $Tr[\rho H]$. It is natural that the variance for a quantum state ρ and an observable H is defined by $V_\rho(H) = Tr[\rho(H - Tr[\rho H]I)^2] = Tr[\rho H^2] - Tr[\rho H]^2$. It is famous that we have

$$V_\rho(A)V_\rho(B) \geq \frac{1}{4}|Tr[\rho[A,B]]|^2 \tag{2.1}$$

for a quantum state ρ and two observables A and B. The further strong results was given by Schrödinger

$$V_\rho(A)V_\rho(B) - |Cov_\rho(A,B)|^2 \geq \frac{1}{4}|Tr[\rho[A,B]]|^2,$$

where the covariance is defined by $Cov_\rho(A,B) = Tr[\rho(A - Tr[\rho A]I)(B - Tr[\rho B]I)]$. However, the uncertainty relation for the Wigner-Yanase skew information failed. (See [9, 5, 11])

$$I_\rho(A)I_\rho(B) \geq \frac{1}{4}|Tr[\rho[A,B]]|^2.$$

Recently, S.Luo introduced the quantity $U_\rho(H)$ representing a quantum uncertainty excluding the classical mixture:

$$U_\rho(H) = \sqrt{V_\rho(H)^2 - (V_\rho(H) - I_\rho(H))^2}, \tag{2.2}$$

then he derived the uncertainty relation on $U_\rho(H)$ in [8]:

$$U_\rho(A)U_\rho(B) \geq \frac{1}{4}|Tr[\rho[A,B]]|^2. \tag{2.3}$$

Note that we have the following relation

$$0 \leq I_\rho(H) \leq U_\rho(H) \leq V_\rho(H). \tag{2.4}$$

The inequality (2.3) is a refinement of the inequality (2.1) in the sense of (2.4). In this section, we study one-parameter extended inequality for the inequality (2.3).

Definition 2.1 *For $0 \leq \alpha \leq 1$, a quantum state ρ and an observable H, we define the Wigner-Yanase-Dyson skew information*

$$\begin{aligned} I_{\rho,\alpha}(H) &= \frac{1}{2}Tr[(i[\rho^\alpha, H_0])(i[\rho^{1-\alpha}, H_0])] \\ &= Tr[\rho H_0^2] - Tr[\rho^\alpha H_0 \rho^{1-\alpha} H_0] \end{aligned} \tag{2.5}$$

and we also define

$$\begin{aligned} J_{\rho,\alpha}(H) &= \frac{1}{2}Tr[\{\rho^\alpha, H_0\}\{\rho^{1-\alpha}, H_0\}] \\ &= Tr[\rho H_0^2] + Tr[\rho^\alpha H_0 \rho^{1-\alpha} H_0], \end{aligned} \tag{2.6}$$

where $H_0 = H - Tr[\rho H]I$ and we denote the anti-commutator by $\{X,Y\} = XY + YX$.

Note that we have

$$\frac{1}{2}Tr[(i[\rho^\alpha, H_0])(i[\rho^{1-\alpha}, H_0])] = \frac{1}{2}Tr[(i[\rho^\alpha, H])(i[\rho^{1-\alpha}, H])]$$

but we have

$$\frac{1}{2}Tr[\{\rho^{\alpha},H_0\}\{\rho^{1-\alpha},H_0\}] \neq \frac{1}{2}Tr[\{\rho^{\alpha},H\}\{\rho^{1-\alpha},H\}].$$

Then we have the following inequalities:

$$I_{\rho,\alpha}(H) \leq I_{\rho}(H) \leq J_{\rho}(H) \leq J_{\rho,\alpha}(H), \tag{2.7}$$

since we have $Tr[\rho^{1/2}H\rho^{1/2}H] \leq Tr[\rho^{\alpha}H\rho^{1-\alpha}H]$. (See [1, 2] for example.) If we define

$$U_{\rho,\alpha}(H) = \sqrt{V_{\rho}(H)^2 - (V_{\rho}(H) - I_{\rho,\alpha}(H))^2}, \tag{2.8}$$

as a direct generalization of Eq.(2.2), then we have

$$0 \leq I_{\rho,\alpha}(H) \leq U_{\rho,\alpha}(H) \leq U_{\rho}(H) \tag{2.9}$$

due to the first inequality of (2.7). We also have

$$U_{\rho,\alpha}(H) = \sqrt{I_{\rho,\alpha}(H)J_{\rho,\alpha}(H)}.$$

From the inequalities (2.4),(2.8),(2.9), our situation is that we have

$$0 \leq I_{\rho,\alpha}(H) \leq I_{\rho}(H) \leq U_{\rho}(H)$$

and

$$0 \leq I_{\rho,\alpha}(H) \leq U_{\rho,\alpha}(H) \leq U_{\rho}(H).$$

Our concern is to show an uncertainty relation with respect to $U_{\rho,\alpha}(H)$ as a direct generalization of the inequality (2.3) such that

$$U_{\rho,\alpha}(A)U_{\rho,\alpha}(B) \geq \frac{1}{4}|Tr[\rho[A,B]]|^2 \tag{2.10}$$

Recently the following theorem was given.

Theorem 2.1 ([6]) *For a quantum state ρ and observables A, B and $0 \leq \alpha \leq 1$,*

$$U_{\rho,\alpha}(A)U_{\rho,\alpha}(B) \geq \frac{1}{4}|Tr[(\rho - \rho^{|2\alpha-1|})[A,B]]|^2. \tag{2.11}$$

However we gave the counter example for (2.11) as follows; Let

$$\rho = \begin{pmatrix} \frac{1}{64} & 0 & 0 \\ 0 & \frac{1}{16} & 0 \\ 0 & 0 & \frac{59}{64} \end{pmatrix}, A = \begin{pmatrix} 0 & i & 0 \\ -i & 0 & 0 \\ 0 & 0 & 0 \end{pmatrix}, B = \begin{pmatrix} 0 & 1 & 0 \\ 1 & 0 & 0 \\ 0 & 0 & 0 \end{pmatrix}, \alpha = \frac{3}{4}.$$

Then we have

$$U_{\rho,\alpha}(A)U_{\rho,\alpha}(B) = 0.00170898\cdots,$$

$$\frac{1}{4}|Tr[(\rho - \rho^{|2\alpha-1|})[A,B]]|^2 = 0.00610351\cdots.$$

Journal of Physics: Conference Series **201** (2010) 012015 doi:10.1088/1742-6596/201/1/012015

3. Main result

We give the main theorem which modifies (2.11) as follows;

Theorem 3.1 *For a quantum state* ρ *and observables* A, B *and* $0 \leq \alpha \leq 1$, *we have*

$$U_{\rho,\alpha}(A)U_{\rho,\alpha}(B) \geq \alpha(1-\alpha)|Tr[\rho[A,B]]|^2. \tag{3.1}$$

We use the several lemmas to prove the theorem 3.1. By spectral decomposition, there exists an orthonormal basis $\{\phi_1, \phi_2, \ldots\}$ consisting of eigenvectors of ρ. Let $\lambda_1, \lambda_2, \ldots$ be the corresponding eigenvalues, where $\sum_{i=1}^{\infty} \lambda_i = 1$ and $\lambda_i \geq 0$. Thus, ρ has a spectral representation

$$\rho = \sum_{i=1}^{\infty} \lambda_i |\phi_i\rangle\langle\phi_i|. \tag{3.2}$$

Lemma 3.1

$$I_{\rho,\alpha}(H) = \sum_{i<j} (\lambda_i + \lambda_j - \lambda_i^{\alpha}\lambda_j^{1-\alpha} - \lambda_i^{1-\alpha}\lambda_j^{\alpha})|\langle\phi_i|H_0|\phi_j\rangle|^2.$$

Proof of Lemma 3.1. By (3.2),

$$\rho H_0^2 = \sum_{i=1}^{\infty} \lambda_i |\phi_i\rangle\langle\phi_i|H_0^2.$$

Then

$$Tr[\rho H_0^2] = \sum_{i=1}^{\infty} \lambda_i \langle\phi_i|H_0^2|\phi_i\rangle = \sum_{i=1}^{\infty} \lambda_i \|H_0|\phi_i\rangle\|^2. \tag{3.3}$$

Since

$$\rho^{\alpha} H_0 = \sum_{i=1}^{\infty} \lambda_i^{\alpha} |\phi_i\rangle\langle\phi_i|H_0$$

and

$$\rho^{1-\alpha} H_0 = \sum_{i=1}^{\infty} \lambda_i^{1-\alpha} |\phi_i\rangle\langle\phi_i|H_0,$$

we have

$$\rho^{\alpha} H_0 \rho^{1-\alpha} H_0 = \sum_{i,j=1}^{\infty} \lambda_i^{\alpha}\lambda_j^{1-\alpha} |\phi_i\rangle\langle\phi_i|H_0|\phi_j\rangle\langle\phi_j|H_0.$$

Thus

$$\begin{aligned} Tr[\rho^{\alpha} H_0 \rho^{1-\alpha} H_0] &= \sum_{i,j=1}^{\infty} \lambda_i^{\alpha}\lambda_j^{1-\alpha} \langle\phi_i|H_0|\phi_j\rangle\langle\phi_j|H_0|\phi_i\rangle \\ &= \sum_{i,j=1}^{\infty} \lambda_i^{\alpha}\lambda_j^{1-\alpha} |\langle\phi_i|H_0|\phi_j\rangle|^2. \end{aligned} \tag{3.4}$$

From (2.5), (3.3), (3.4),

$$\begin{aligned} I_{\rho,\alpha}(H) &= \sum_{i=1}^{\infty} \lambda_i \| H_0|\phi_i\rangle\|^2 - \sum_{i,j=1}^{\infty} \lambda_i^{\alpha}\lambda_j^{1-\alpha}|\langle\phi_i|H_0|\phi_j\rangle|^2 \\ &= \sum_{i,j=1}^{\infty} (\lambda_i - \lambda_i^{\alpha}\lambda_j^{1-\alpha})|\langle\phi_i|H_0|\phi_j\rangle|^2 \\ &= \sum_{i<j} (\lambda_i + \lambda_j - \lambda_i^{\alpha}\lambda_j^{1-\alpha} - \lambda_i^{1-\alpha}\lambda_j^{\alpha})|\langle\phi_i|H_0|\phi_j\rangle|^2. \end{aligned}$$

□

Lemma 3.2

$$J_{\rho,\alpha}(H) \geq \sum_{i<j} (\lambda_i + \lambda_j + \lambda_i^{\alpha}\lambda_j^{1-\alpha} + \lambda_i^{1-\alpha}\lambda_j^{\alpha})|\langle\phi_i|H_0|\phi_j\rangle|^2.$$

Proof of Lemma 3.2. By (2.6), (3.3), (3.4), we have

$$\begin{aligned} J_{\rho,\alpha}(H) &= \sum_{i=1}^{\infty} \lambda_i \| H_0|\phi_i\rangle\|^2 + \sum_{i,j=1}^{\infty} \lambda_i^{\alpha}\lambda_j^{1-\alpha}|\langle\phi_i|H_0|\phi_j\rangle|^2 \\ &= \sum_{i,j=1}^{\infty} (\lambda_i + \lambda_i^{\alpha}\lambda_j^{1-\alpha})|\langle\phi_i|H_0|\phi_j\rangle|^2 \\ &= 2\sum_{i=1}^{\infty} \lambda_i|\langle\phi_i|H_0|\phi_i\rangle|^2 + \sum_{i\neq j} (\lambda_i + \lambda_i^{\alpha}\lambda_j^{1-\alpha})|\langle\phi_i|H_0|\phi_j\rangle|^2 \\ &= 2\sum_{i=1}^{\infty} \lambda_i|\langle\phi_i|H_0|\phi_i\rangle|^2 + \sum_{i<j} (\lambda_i + \lambda_j + \lambda_i^{\alpha}\lambda_j^{1-\alpha} + \lambda_i^{1-\alpha}\lambda_j^{\alpha})|\langle\phi_i|H_0|\phi_j\rangle|^2 \\ &\geq \sum_{i<j} (\lambda_i + \lambda_j + \lambda_i^{\alpha}\lambda_j^{1-\alpha} + \lambda_i^{1-\alpha}\lambda_j^{\alpha})|\langle\phi_i|H_0|\phi_j\rangle|^2. \end{aligned}$$

□

Lemma 3.3 *For any $t > 0$ and $0 \leq \alpha \leq 1$, the following inequality holds;*

$$(1-2\alpha)^2(t-1)^2 - (t^{\alpha} - t^{1-\alpha})^2 \geq 0. \tag{3.5}$$

Proof of Lemma 3.3. If $\alpha = 0$ or $\frac{1}{2}$ or 1, then it is clear that (3.5) is satisfied. Now we put

$$F(t) = (1-2\alpha)^2(t-1)^2 - (t^{\alpha} - t^{1-\alpha})^2.$$

We have

$$F'(t) = 2(1-2\alpha)^2 t - 2\alpha t^{2\alpha-1} - 2(1-\alpha)t^{1-2\alpha} + 8\alpha(1-\alpha).$$

And we also have

$$F^{''}(t) = 2(1-2\alpha)^2 - 2\alpha(2\alpha-1)t^{2\alpha-2} - 2(1-\alpha)(1-2\alpha)t^{-2\alpha}$$

and

$$\begin{aligned} & F^{'''}(t) \\ = & \; 4\alpha(1-2\alpha)(1-\alpha)t^{-2\alpha-1} - 4\alpha(1-2\alpha)(1-\alpha)t^{2\alpha-3} \\ = & \; 4\alpha(1-2\alpha)(1-\alpha)\left(\frac{1}{t^{1+2\alpha}} - \frac{1}{t^{3-2\alpha}}\right). \end{aligned}$$

If $\frac{1}{2} < \alpha < 1$, then $1+2\alpha > 3-2\alpha$. Then it is easy to show that $F^{'''}(t) < 0$ for $t < 1$ and $F^{'''}(t) > 0$ for $t > 1$. On the other hand if $0 < \alpha < \frac{1}{2}$, then $1+2\alpha < 3-2\alpha$. Then it is easy to show that $F^{'''}(t) < 0$ for $t < 1$ and $F^{'''}(t) > 0$ for $t > 1$. Since $F^{''}(1) = 0$, we can get $F^{''}(t) > 0$. Since $F^{'}(1) = 0$, we also have $F^{'}(t) < 0$ for $t < 1$ and $F^{'}(t) > 0$ for $t > 1$. Since $F(1) = 0$, we finally get $F(t) \geq 0$ for all $t > 0$. Therefore we have (3.5). □

Proof of Theorem 3.1. We put $t = \dfrac{\lambda_i}{\lambda_j}$ in (3.5). Then we have

$$(1-2\alpha)^2\left(\frac{\lambda_i}{\lambda_j} - 1\right)^2 - \left(\left(\frac{\lambda_i}{\lambda_j}\right)^\alpha - \left(\frac{\lambda_i}{\lambda_j}\right)^{1-\alpha}\right)^2 \geq 0.$$

And we get

$$(1-2\alpha)^2(\lambda_i - \lambda_j)^2 - (\lambda_i^\alpha\lambda_j^{1-\alpha} - \lambda_i^{1-\alpha}\lambda_j^\alpha)^2 \geq 0$$

and

$$(\lambda_i - \lambda_j)^2 - (\lambda_i^\alpha\lambda_j^{1-\alpha} - \lambda_i^{1-\alpha}\lambda_j^\alpha)^2 \geq 4\alpha(1-\alpha)(\lambda_i - \lambda_j)^2$$

and

$$(\lambda_i + \lambda_j)^2 - (\lambda_i^\alpha\lambda_j^{1-\alpha} + \lambda_i^{1-\alpha}\lambda_j^\alpha)^2 \geq 4\alpha(1-\alpha)(\lambda_i - \lambda_j)^2. \tag{3.6}$$

Since

$$\begin{aligned} Tr[\rho[A,B]] &= Tr[\rho[A_0, B_0]] \\ &= 2iImTr[\rho A_0 B_0] \\ &= 2iIm\sum_{i<j}(\lambda_i - \lambda_j)\langle\phi_i|A_0|\phi_j\rangle\langle\phi_j|B_0|\phi_i\rangle \\ &= 2i\sum_{i<j}(\lambda_i - \lambda_j)Im\langle\phi_i|A_0|\phi_j\rangle\langle\phi_j|B_0|\phi_i\rangle, \end{aligned}$$

$$\begin{aligned} |Tr[\rho[A,B]]| &= 2|\sum_{i<j}(\lambda_i - \lambda_j)Im\langle\phi_i|A_0|\phi_j\rangle\langle\phi_j|B_0|\phi_i\rangle| \\ &\leq 2\sum_{i<j}|\lambda_i - \lambda_j||Im\langle\phi_i|A_0|\phi_j\rangle\langle\phi_j|B_0|\phi_i\rangle|. \end{aligned}$$

Then we have

$$|Tr[\rho[A,B]]|^2 \leq 4\{\sum_{i<j}|\lambda_i - \lambda_j||Im\langle\phi_i|A_0|\phi_j\rangle\langle\phi_j|_0|\phi_i\rangle|\}^2.$$

By (3.6) and Schwarz inequality,

$$\begin{aligned}
& \alpha(1-\alpha)|Tr[\rho[A,B]]|^2 \\
\leq\ & 4\alpha(1-\alpha)\{\sum_{i<j}|\lambda_i-\lambda_j||Im\langle\phi_i|A_0|\phi_j\rangle\langle\phi_j|B_0|\phi_i\rangle|\}^2 \\
=\ & \{\sum_{i<j}2\sqrt{\alpha(1-\alpha)}|\lambda_i-\lambda_j||Im\langle\phi_i|A_0|\phi_j\rangle\langle\phi_j|B_0|\phi_i\rangle|\}^2 \\
\leq\ & \{\sum_{i<j}2\sqrt{\alpha(1-\alpha)}|\lambda_i-\lambda_j||\langle\phi_i|A_0|\phi_j\rangle||\langle\phi_j|B_0|\phi_i\rangle|\}^2 \\
\leq\ & \{\sum_{i<j}\{(\lambda_i+\lambda_j)^2-(\lambda_i^{\alpha}\lambda_j^{1-\alpha}+\lambda_i^{1-\alpha}\lambda_j^{\alpha})^2\}^{1/2}|\langle\phi_i|A_0|\phi_j\rangle||\langle\phi_j|B_0|\phi_i\rangle|\}^2 \\
\leq\ & \sum_{i<j}(\lambda_i+\lambda_j-\lambda_i^{\alpha}\lambda_j^{1-\alpha}-\lambda_i^{1-\alpha}\lambda_j^{\alpha})|\langle\phi_i|A_0|\phi_j\rangle|^2 \\
& \times\sum_{i<j}(\lambda_i+\lambda_j+\lambda_i^{\alpha}\lambda_j^{1-\alpha}+\lambda_i^{1-\alpha}\lambda_j^{\alpha})|\langle\phi_i|B_0|\phi_j\rangle|^2.
\end{aligned}$$

Then we have

$$I_{\rho,\alpha}(A)J_{\rho,\alpha}(B)\geq\alpha(1-\alpha)|Tr[\rho[A,B]]|^2.$$

We also have

$$I_{\rho,\alpha}(B)J_{\rho,\alpha}(A)\geq\alpha(1-\alpha)|Tr[\rho[A,B]]|^2.$$

Hence we have the final result (3.1). □

Remark 3.1 *We remark that (2.3) is derived by putting $\alpha=1/2$ in (3.1). Then Theorem 3.1 is a generalization of the result of Luo [8].*

Remark 3.2 *We remark that Conjecture 2.3 in [3] does not hold in general. The Conjecture is (2.10). A counterexample is given as follows. Let*

$$\rho=\begin{pmatrix}3/4 & 0\\ 0 & 1/4\end{pmatrix},\ A=\begin{pmatrix}0 & i\\ -i & 0\end{pmatrix},\ B=\begin{pmatrix}0 & 1\\ 1 & 0\end{pmatrix},\ \alpha=1/3.$$

We have

$$I_{\rho,\alpha}(A)J_{\rho,\alpha}(B)=I_{\rho,\alpha}(B)I_{\rho,\alpha}(A)=0.22457296\cdots$$

and $|Tr[\rho[A,B]]|^2=1$. These imply

$$U_{\rho,\alpha}(A)U_{\rho,\alpha}(B)=0.22457296\cdots<\frac{1}{4}|Tr[\rho[A,B]]|^2=0.25.$$

On the other hand we have

$$U_{\rho,\alpha}(A)U_{\rho,\alpha}(B)>\alpha(1-\alpha)|Tr[\rho[A,B]]|^2=0.2222222\cdots.$$

We also give a counterexample for Conjecture 2.10 in [3]. The inequality

$$U_{\rho,\alpha}(A)U_{\rho,\alpha}(B)\geq\frac{1}{4}|Tr[(\frac{\rho^{\alpha}+\rho^{1-\alpha}}{2})^2[A,B]]|^2$$

is not correct in general, because $LHS=0.22457296\cdots$, $RHS=0.23828105995\cdots$.

Remark 3.3 *There seems to be a similarity in mathematical formulations between the generalization of $\rho^{1/2}$ to ρ^{α} in Wigner-Yanase-Dyson skew information and ρ^q in the generalized entropies such as Renyi or Tsallis entropy.*

Journal of Physics: Conference Series **201** (2010) 012015 doi:10.1088/1742-6596/201/1/012015

Acknowledgments
This research was partially supported by the Ministry of Education, Science, Sports and Culture, Grant-in-Aid for Scientific Research (B), 18300003 and (C), 20540175

References
[1] J.C.Bourin, *Some inequalities for norms on matrices and operators*, Linear Algebra and its Applications, vol.292(1999), pp.139-154.

[2] J.I.Fujii, *A trace inequality arising from quantum information theory*, Linear Algebra and its Applications, vol.400(2005), pp.141-146.

[3] S.Furuichi, K.Yanagi and K.Kuriyama, *Trace inequalities on a generalized Wigner-Yanase skew information*, J. Math. Anal. Appl., vol.356(2009), pp.179-185.

[4] W.Heisenberg, *Über den anschaulichen Inhat der quantummechanischen Kinematik und Mechanik*, Zeitschrift für Physik, vol.43(1927), pp.172-198.

[5] H.Kosaki, *Matrix trace inequality related to uncertainty principle*, Internatonal Journal of Mathematics, vol.16(2005), pp.629-646.

[6] D.Li, X.Li, F.Wang, H.Huang, X.Li and L.C.Kwek, *Uncertainty relation of mixed states by means of Wigner-Yanase-Dyson information*, Physical Review A, vol.79(2009), pp.052106-1-4.

[7] E.H.Lieb, *Convex trace functions and the Wigner-Yanase-Dyson conjecture*, Adv. Math., vol.11(1973), pp.267-288.

[8] S.Luo, *Heisenberg uncertainty relation for mixed states*, Phys. Rev. A, vol.72(2005), p.042110.

[9] S.Luo and Q.Zhang, *On skew information*, IEEE Trans. Information Theory, vol.50(2004), pp.1778-1782, and *Correction to "On skew information"*, IEEE Trans. Information Theory, vol.51(2005), p.4432.

[10] E.P.Wigner and M.M.Yanase, *Information content of distribution*, Proc. Nat. Acad. Sci. U,S,A., vol.49(1963), pp.910-918.

[11] K.Yanagi, S.Furuichi and K.Kuriyama, *A generalized skew information and uncertainty relation*, IEEE Trans. Information Theory, vol.51(2005), pp.4401-4404.

[12] K.Yanagi, *Uncertainty relation on Wigner-Yanase-Dyson skew information*, arXiv:0907.1576v1[quant-ph] 9 Jul 2009.

Journal of Physics: Conference Series **201** (2010) 012016 doi:10.1088/1742-6596/201/1/012016

On generalized Fisher informations and Cremér-Rao type inequalities

Shigeru Furuichi

Department of Computer Science and System Analysis, College of Humanities and Sciences, Nihon University, 3-25-40, Sakurajyousui, Setagaya-ku, Tokyo, 156-8550, Japan

E-mail: `furuichi@chs.nihon-u.ac.jp`

Abstract. We define q-Fisher informations and show q-Cramér-Rao type inequalities that the q-Gaussian distribution with special q-variances attains the minimum value of the q-Fisher informations.

1. Introduction

The maximum entropy principle is fundamental in statistical physics. In nonextensive statistical physics, the maximum Tsallis entropy principle is also important and has been studied by many literatures [17, 18, 10, 19, 3, 1, 7]. The maximum Tsallis entropy principle states that Tsallis distribution (q-Gaussian distribution) attains the maximum value of the Tsallis entropy, subject to the constraint on the q-expectation value and the q-variance. On the other hand, it is well known that the Gaussian distribution maximize the entropy and minimize the Fisher information (Cramér-Rao inequality) [4]. Therefore it is quite natural that we analogically have the following question: q-Gaussian minimizes q-Fisher information? We here consider it for two cases in nonextensive statistical physics. This brief report is based on our previous paper [7] so that we omit the proofs.

2. The q-Fisher information based on the q-expectation value

We denote the q-logarithmic function $\ln_q$ by

$$\ln_q(x) \equiv \frac{x^{1-q}-1}{1-q}, \qquad (q \in \mathbb{R}, q \neq 1, x > 0).$$

In this section, we give a definition of the q-Fisher information, based on the q-expectation value E_q.

Definition 1 *For random variable X with probability density function $f(x)$, we define the q-score function $s_q(x)$ and the q-Fisher information $J_q(X)$ by*

$$s_q(x) \equiv \frac{d \ln_q f(x)}{dx}, \tag{1}$$

$$J_q(X) \equiv E_q\left[s_q(X)^2\right], \tag{2}$$

where the q-expectation value E_q is defined by $E_q[g(X)] \equiv \int_{-\infty}^{\infty} f(x)^q g(x)dx$ for random variables $g(X)$ for any continuous function $g(x)$ and the probability density function $f(x)$.

Then we have the following theorem.

Theorem 1 *(q-Cramér-Rao type inequality [7]) For given the q-expectation value $\mu_q \equiv E_q[X]$, the q-variance $\sigma_q^2 \equiv E_q\left[(X-\mu_q)^2\right]$ and the random variables X with the probability density function $p(x)$ we have the following q-Cramér-Rao type inequality:*

$$J_q(X) \geq \frac{1}{\sigma_q^2} \quad for \;\; q \in [0,1) \cup (1,3). \tag{3}$$

In addition, the equality holds if the probability density function is given by the q-Gaussian density function:

$$\phi_q(x) \equiv \frac{1}{Z_q} \exp_q\left\{-\frac{\beta_q (x-\mu_q)^2}{\sigma_q^2}\right\}, \;\; Z_q \equiv \int_{-\infty}^{\infty} \exp_q\left\{-\frac{\beta_q (x-\mu_q)^2}{\sigma_q^2}\right\}, \;\; \beta_q \equiv \frac{1}{3-q},$$

with the q-variance:

$$\sigma_q = \frac{2^{\frac{q}{1-q}} (3-q)^{\frac{q+1}{2(q-1)}} (1-q)^{\frac{1}{2}}}{B\left(\frac{1}{2}, \frac{1}{1-q}\right)}, \quad (0 \leq q < 1) \tag{4}$$

or

$$\sigma_q = \frac{2^{\frac{1}{1-q}} (3-q)^{\frac{3-q}{2(q-1)}} (q-1)^{\frac{1}{2}}}{B\left(\frac{1}{q-1} - \frac{1}{2}, \frac{1}{2}\right)}, \quad (1 < q < 3). \tag{5}$$

3. The q-Fisher information based on the normalized q-expectation value

In this section, we give another definition of q-Fisher information, based on the normalized q-expectation value $E_q^{(nor)}$.

Definition 2 *For the random variable X with the probability density function $f(x)$, we define the q-score function $s_q(x)$ and the q-Fisher information $J_q(X)$ by*

$$s_q(x) \equiv \frac{d \ln_q f(x)}{dx}, \tag{6}$$

$$J_q^{(nor)}(X) \equiv E_q^{(nor)}\left[s_q(X)^2\right], \tag{7}$$

where the normalized q-expectation value $E_q^{(nor)}$ is defined by $E_q^{(nor)}[g(X)] \equiv \frac{\int g(x) f(x)^q dx}{\int f(x)^q dx}$ for random variables $g(X)$ for any continuous function $g(x)$ and the probability density function $f(x)$.

Note that our q-Fisher informations defined in Definition 1 and Definition 2 are different from those in several literature [13, 12, 2, 11, 15, 14, 8]. Then we have the following theorem.

Theorem 2 *(q-Cramér-Rao type inequality [7]) Given the random variable X with the probability density function $p(x)$, the q-expectation value $\mu_q \equiv E_q^{(nor)}[X]$ and the q-variance $\sigma_q^2 \equiv E_q^{(nor)}\left[(X-\mu_q)^2\right]$, we have the following q-Cramér-Rao type inequality:*

$$J_q^{(nor)}(X) \geq \frac{1}{\sigma_q^2}\left(\frac{2}{\int p(x)^q dx} - 1\right) \quad for \;\; q \in [0,1) \cup (1,3). \tag{8}$$

Immediately we have

$$J_q^{(nor)}(X) \geq \frac{1}{\sigma_q^2} \quad for \quad q \in (1,3). \tag{9}$$

In addition, the equality holds if the probability density function is given by the q-Gaussian density function:

$$\phi_q(x) \equiv \frac{1}{Z_q} \exp_q\left\{-\frac{\beta_q (x-\mu_q)^2}{\sigma_q^2}\right\}, \; Z_q \equiv \int_{-\infty}^{\infty} \exp_q\left\{-\frac{\beta_q (x-\mu_q)^2}{\sigma_q^2}\right\}, \; \beta_q \equiv \frac{1}{3-q},$$

with the q-variance:

$$\sigma_q = \frac{2^{\frac{q}{1-q}} (3-q)^{\frac{q+1}{2(q-1)}} (1-q)^{\frac{1}{2}}}{B\left(\frac{1}{2}, \frac{1}{1-q}\right)}, \quad (0 \leq q < 1) \tag{10}$$

or

$$\sigma_q = \frac{2^{\frac{1}{1-q}} (3-q)^{\frac{3-q}{2(q-1)}} (q-1)^{\frac{1}{2}}}{B\left(\frac{1}{q-1} - \frac{1}{2}, \frac{1}{2}\right)}, \quad (1 < q < 3). \tag{11}$$

The q-variances σ_q given in the above, Eq.(10) and Eq.(11) are same to those in Eq.(4) and Eq.(5) . The proof is given in [7] in an analogical way to that in [9]. Both theorems in section 2 and section 3 recover the usual Cramér-Rao inequality $J_1(X) \geq \frac{1}{\sigma_1^2}$ ($J_1^{(nor)}(X) \geq \frac{1}{\sigma_1^2}$) in the limit $q \to 1$.

4. On an example of the q-Fisher informations J_q and $J_q^{(nor)}$ and the feature of the q-variance σ_q appeared in the previous sections

Here, we give the typical example for two q-Fisher informations. For the random variable G obeying the q-Gaussian distribution

$$\phi_q(x) \equiv \frac{1}{Z_q} \exp_q\left\{-\frac{\beta_q (x-\mu_q)^2}{\sigma_q^2}\right\},$$

where $\beta_q \equiv \frac{1}{3-q}$ and the q-partition function $Z_q \equiv \int_{-\infty}^{\infty} \exp_q\left\{-\frac{\beta_q (x-\mu_q)^2}{\sigma_q^2}\right\} dx$, the q-score function is calculated as

$$s_q(x) = -\frac{2\beta_q Z_q^{q-1}}{\sigma_q^2}(x-\mu_q),$$

which is linear with respect to x. Thus we can calculate the q-Fisher informations as

$$J_q(G) = J_q^{(nor)}(G) = \frac{4\beta_q^2 Z_q^{2q-2}}{\sigma_q^2}. \tag{12}$$

Note that we have

$$\lim_{q\to 1} J_q(G) = \lim_{q\to 1} J_q^{(nor)}(G) = \frac{1}{\sigma_1^2}. \tag{13}$$

Finally we give comments on σ_q. We firstly note on the limit $q \to 1$ for the q-variances σ_q given in Eq.(4) and Eq.(5) (or Eq.(10) and Eq.(11)). The following results were checked by the computer software:

$$\lim_{q\to 1-0} \sigma_q = \lim_{q\to 1-0} \frac{2^{\frac{q}{1-q}} (3-q)^{\frac{q+1}{2(q-1)}} (1-q)^{\frac{1}{2}}}{B\left(\frac{1}{2}, \frac{1}{1-q}\right)} = \lim_{r\to +0} \frac{2^{\frac{1-r}{r}} (2+r)^{\frac{r-2}{2r}} r^{\frac{1}{2}}}{B\left(\frac{1}{2}, \frac{1}{r}\right)} = \frac{1}{\sqrt{2e\pi}}$$

and

$$\lim_{q\to 1+0}\sigma_q = \lim_{q\to 1+0}\frac{2^{\frac{1}{1-q}}(3-q)^{\frac{3-q}{2(q-1)}}(q-1)^{\frac{1}{2}}}{B\left(\frac{1}{q-1}-\frac{1}{2},\frac{1}{2}\right)} = \frac{1}{\sqrt{2e\pi}}.$$

Furthermore, the following figure represents σ_q for $q \in [0,3)$.

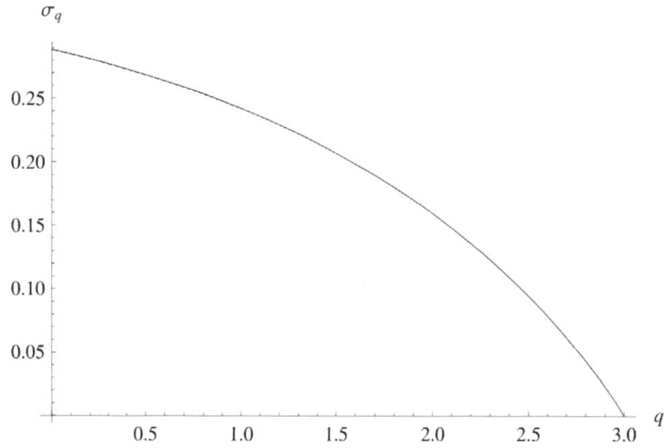

Figure 1. The q-variances σ_q given in Eq.(4) and Eq.(5) (or Eq.(10) and Eq.(11))

5. Conclusion

In our previous paper [6], we gave the rough meaning of the parameter q from the information-theoretical viewpoint. In [6], we showed that the Tsallis entropies for $q \geq 1$ had the subadditivity and therefore we had several information-theoretical properties in the case of $q \geq 1$. However, the Tsallis entropies for $q < 1$ did not have such properties. As similar as the case of the Tsallis entropies, in Theorem 2 we have found that the q-Fisher information have the quite same situation such that we have $J_q^{(nor)}(X) \geq \frac{1}{\sigma_q^2}$ for $q \geq 1$, however for the case of $q < 1$, we do not have any relation between $J_q^{(nor)}(X)$ and $\frac{1}{\sigma_q^2}$ other than the inequality (8). Therefore these results give us the difference of the q-Fisher information $J_q^{(nor)}(X)$ for $q \in [0,1)$ and $J_q^{(nor)}(X)$ for $q \in (1,3)$, as similar as the Tsallis entropies did in [6]. Summarizing these results, we may conclude that the Tsallis entropies and the q-Fisher information $J_q^{(nor)}(X)$ based on the normalized q-expectation value, make a sense for the case of $q \geq 1$ in our setting. As for the q-Fisher information $J_q(X)$ based on the q-expectation value, we do not find such difference on Cramér-Rao type inequality.

We close this section giving a comment on a possible application of our q-Fisher informations. The famous central limit theorem in probability theory, shows the distribution function of the standardized sum of an independent sequence of random variables convergences to Gaussian distribution under a certain assumpution. The classical central limit theorem is usually proved by the characteristic function. However we have another method of its proof. Namely, the Fisher information can be applied to prove the classical central limit theorem [5, 16, 9]. Recently, the q-central limit theorem for $q \geq 1$ was proved in [20] by introducing new notions such as the q-independence, the q-Fourier transformation and the q-characteristic function and so on. Therefore we may expect that a new proof of the q-central limit theorem may be given by applying the q-Fisher information in the future.

Acknowledgement
The author was partially supported by the Japanese Ministry of Education, Science, Sports and Culture, Grant-in-Aid for Scientific Research (B), 18300003 and Grant-in-Aid for Encouragement of Young Scientists (B), 20740067.

References

[1] S.Abe, Heat and entropy in nonextensive thermodynamics: transmutation from Tsallis theory to Rényi-entropy-based theory,Physica A,Vol.300(2001),pp.417-423.
[2] S. Abe, Geometry of escort distributions,Phys. Rev. E, Vol.68(2003), 031101.
[3] S.Abe, S. Martńez, F. Pennini, and A. Plastino, Nonextensive thermodynamic relations, Phys.Lett.A,Vol.281(2001),pp.126-130.
[4] S.Amari and H.Nagaoka, Methods of Information Geometry, OXFORD UNIV. PRESS, 2000.
[5] L.D.Brown, A proof of the central limit theorem motivated by the Cramér-Rao inequality, G.Kallianpur, P.R.Krishnaiah and J.K.Ghosh eds., Statistics and Probability: Essays in Honor of C.R.Rao, North-Holland Publishing Campany (1982),pp.141-148.
[6] S.Furuichi, Information theoretical properties of Tsallis entropies, J. Math. Phys., Vol.47(2006), 023302.
[7] S.Furuichi, On the maximum entropy principle and the minimization of the Fisher information in Tsallis statistics, J. Math. Phys., Vol.50(2009),013303.
[8] H. Hasegawa, Stationary and dynamical properties of information entropies in nonextensive systems,Phys. Rev. E,Vol.77(2008),031133.
[9] O.Johnson, Information theory and the central limit theorem, Imperial College Press, 2004.
[10] S. Martinez, F. Nicolás, F. Pennini and A. Plastino, Tsallis' entropy maximization procedure revisited, Physica A,Vol.286(2000), pp.489-502.
[11] F. Pennini and A. Plastino, Power-law distributions and Fisher's information measure, Physica A, Vol.334(2004), pp.132-138.
[12] F. Pennini, A.R. Plastino and A. Plastino, Renyi entropies and Fisher informations as measures of nonextensivity in a Tsallis setting, Physica A, Vol.258 (1998), pp.446-457.
[13] A. Plastino, A. R. Plastino and H. G. Miller,Tsallis nonextensive thermostatistics and Fisher's information measure, Physica A,Vol.235(1997),pp.557-588.
[14] M. Portesi, F. Pennini and A. Plastino,Geometrical aspects of a generalized statistical mechanics, Physica A, Vol.373(2007), pp.273-282.
[15] M. Portesi, A. Plastino, and F. Pennini, Information measures based on Tsallis' entropy and geometric considerations for thermodynamic systems, Physica A, Vol.365(2006),pp.173-176.
[16] R.Shimizu, On Fisher's amount of information for location family, G.P.Patil et.al.(eds.),Statistical Distributions in Scientific Work, Vol.3, D.Reidel Publishing Company, Dordrecht-Holland(1975),pp.305-312.
[17] C.Tsallis, Possible generalization of Bolzmann-Gibbs statistics, J.Stat.Phys.,Vol.52(1988),pp.479-487.
[18] C. Tsallis et al., Nonextensive Statistical Mechanics and Its Applications, edited by S. Abe and Y. Okamoto (Springer-Verlag, Heidelberg,2001); see also the comprehensive list of references at http://tsallis.cat.cbpf.br/biblio.htm.
[19] C.Tsallis, R.S.Mendes and A.R.Plastino, The role of constraints within generalized nonextensive statistics, Physica A,Vol.261,(1998),pp.534-554.
[20] S.Umarov,C.Tsallis and S.Steinberg, On a q-central limit theorem consistent with nonextensive statistical mechanics, Milan Journal of Mathematics, Vol.76(2008),pp.307-328.

Mathematical Aspects of Generalized Entropies and their Applications IOP Publishing
Journal of Physics: Conference Series **201** (2010) 012017 doi:10.1088/1742-6596/201/1/012017

Meaning of Non-Extensive Entropies in Micro-Macro Duality

Izumi Ojima
RIMS, Kyoto University, Japan
E-mail: ojima@kurims.kyoto-u.ac.jp

Abstract. The purpose of this presentation is to clarify the universal meaning of generalized non-extensive entropies and/or statistics based on Tsallis entropy from the viewpoint of "*Micro-Macro duality*". What plays here the most important roles is the notions of *scales* and *scaling transformations* which will be seen to be related closely to the main themes of this workshop, generalized non-extensive entropies and/or statistics and the information-geometrical structures of the classifying spaces.

1. Non-additive Statistics & Micro-Macro Duality

Since the importance of *non-additive statistics* does not seem to be fully recognized yet in the science community, I originally intended to clarify in my talk the universal meaning of non-additive statistics based on Tsallis entropy [1, 2], from my own viewpoint of "*Micro-Macro duality*" [3]. After listening to many excellent lectures in this workshop given by the famous experts in the field including Prof. Tsallis himself, however, I have realized that it is redundant to (re-)interpret its essence from some general viewpoints. I have, therefore, decided to shift the focus of my talk into a challenge to reformulate the Tsallis entropy as a solution to an *inverse problem*, which is one of the typical questions arising from the context of Micro-Macro duality. When I gave my talk at the workshop in July 2009, however, my understanding was not sufficient at the most important points in the problem of specifying the context for the Tsallis entropy and its relative version to be uniquely determined. While it is not completely solved yet, there has been good progress after the workshop, which will be reported at the end of this paper.

In this context the points to be discussed are as follows:

1) The essence of "statistics" in general found in the ***relations between "unknown" & "known"*** regions in the world:
While such identifications as unknown $\simeq$ microscopic & known $\simeq$ macroscopic need not be fixed (as opposite connections possible), we can gain a deep insight from analyzing *their mutual relations*. The general scheme based on "Micro-Macro duality" seems to be well suited for this purpose [$\Rightarrow$ Sec.2 & 4].

2) In this context, *scales* and *scale transformations* play essential roles in *moving from one side to another*:
This is also one of the most important aspects treated in "Micro-Macro duality" [$\Rightarrow$ Sec.5].

3) The *distinction and mutual relations* between Micro and Macro, and between qunatum and classical, can be systematically formulated and treated in the above scheme, via the method *to construct a composite system to unify both extreme ends into one* [$\Rightarrow$ Sec.3]. This machinery

seems to be useful for understanding the relations between usual statistics with extensivity and non-extensive one.

4) Important roles of "***inverse problem***" [⇒ Sec.4 & 5]: because of the ***bi-directionality*** inherent in ***duality***, many of seemingly one-directional processes can be inverted, as seen below. I shall try to put Tsallis entropy in the context.

2. Micro-Macro Duality & "Duheme-Quine Paradox"

To discuss the above problems, we explain briefly the essence of *Micro-Macro duality*. It provides a general framework for treating unknown systems and phenomena on the basis of "***dualities***" between *unknown and/or invisible* worlds to be described (symbolically called "Micro") and *known and/or visible* ones to serve as *reference frames* (symbolically "Macro") in the description:

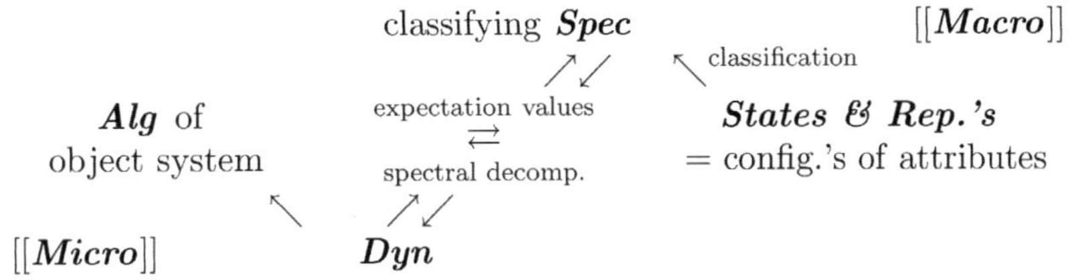

Basic ingredients of our *quadrality scheme* [4] for description, consist of:
Dyn = dynamics (= time development) of object system,
Alg = algebra $\mathcal{A}$ of observables to characterize the object system,
Rep.'s & States = states and representations of $\mathcal{A}$ to describe configurations of the system via measured values,
Spec = classifying space to parametrize the classified structures of the object system and phenomena.

1) Basic duality relation of [Alg ⇄ Reps] is due to the "duality between objects and their attribututes":

[an object = family of attributes] $\overset{\text{dual}}{\rightleftarrows}$ [an attribute= objects sharing that attribute = a set (in mathematics)]

2) The above duality relations themselves change chronologically, in contrast to ***invariant*** quantities
⟹ Duality of [Dyn ⇄ Spec] is based on the "duality between changes and invariants"

3) "5W1H" nature inherent in all the theoretical descriptions of phenomena:

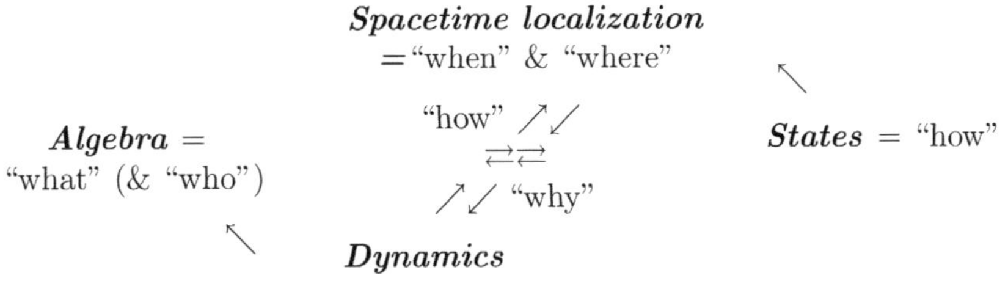

With suitable re-interpretations, many examples of the above scheme can be found in the basic structures of physical (and statistical) theories.

A *No-Go theorem*, "***Duheme-Quine paradox***", however, prevents us from determining uniquely a theory to reproduce phenomenological data, owing to *inevitable finiteness* in *number* and *resolutions* of measured quantities.

$\Longrightarrow$ This *universal dilemma is overcome* in our quadrality scheme in such a way that the very origin of the difficulty leads to a solution of the problem!!: by the necessity to restrict the *aspects to be focused* and the *degrees of exactitude*, a certain *context* is always (pre-)chosen, to within which a theoretical explanation can be unicified by "***Micro-Macro duality***" as a context-dependent *matching condition* [4] between the phenomena to be described and the theory to describe; its mathematical formulation entitles "Macro" as standard reference system satisfying "*universality*".

3. Quantum-Classical Composite System

To explain the above, we define the notion of a ***sector=*** (thermodynamic) ***pure phase***, parametrized by an ***order parameter***:

Mathematically, a ***sector*** $\overset{\text{def}}{=}$ a quasi-equivalence class of ***factor*** states (associated with a facor representation π) of algebra $\mathcal{A}$ of observables, as a *minimal unit* of representations classified by ***quasi-equivalence*** (= unitary equivalence *up to multiplicities*) and characterized by *trivial centre* $\pi(\mathcal{A})'' \cap \pi(\mathcal{A})' =: \mathfrak{z}_\pi(\mathcal{A}) = \mathbb{C}1$ (i.e., *central* observables commuting with all other observables are only scalars $\in \mathbb{C}1$), where the symbol $\mathcal{B}'$ for any subset $\mathcal{B}$ of $\mathcal{A}$ means the *commutant* $\mathcal{B}' := \{T; TB = BT \text{ for } \forall B \in \mathcal{B}\}$ of $\mathcal{B}$. It can be shown that any two factor representations π_1 and π_2 are either mutually quasi-equivalent or ***disjoint*** in the sense that any intertwiner T between π_1 and π_2, $T\pi_1(A) = \pi_2(A)T$ for $\forall A \in \mathcal{A}$, is vanishing: $T = 0$.

Remark: In quantum mechanics (QM) with *finite degrees of freedom*, we have only one sector(!!)$\cdots(*)$, as all the pure states/ representations are *unitarily equivalent* according to Dirac theory of unitary transformations based on Stone-von Neumann uniqueness theorem. This means that the usual QM with finite degrees of freedom is destined, from the beginning, to *fail* in explaining the mutual relations between quantum and classical levels in nature.

The case opposite to this one-*sector* (= *pure phase*) situation is a *mixed phase* $\pi = \bigoplus_{\gamma \in Sp(\mathfrak{z}_\pi(\mathcal{A}))} \pi_\gamma$ consisting of *many sectors* π_γ (= macroscopically different configurations of "Micro") parametrized by *order parameters* as *spectrum* $\gamma \in Sp(\mathfrak{z}_\pi(\mathcal{A}))$ of the non-trivial centre $\mathfrak{z}_\pi(\mathcal{A})$ (= Macro), which is simultaneously diagonalized owing to its commutativity. Thus, we encounter in this situation a ***Quantum-Classical Composite System*** consisting of both ***intra-sectorial*** quantum aspects within each sector and ***inter-sectorial*** classical ones extending over different sectors:

$\diagdown$ inter- $\diagdown$ sector	$\longleftarrow$	$\cdots$	visible = classical	$\cdots$	$\longrightarrow$
intra- $\diagdown$ sector $\diagdown$	γ_1	γ_2	***names*** γ ***of sectors*** $\in Sp(\mathfrak{z}_\pi)$	γ_N	$\cdots$
inside of ***sectors*** $\uparrow$	$\vdots$	$\vdots$	$\vdots$	$\vdots$	$\vdots$
$\parallel$	$\pi_{\gamma_1}\vdots$	$\pi_{\gamma_2}\vdots$	$\pi_\gamma\vdots$	$\pi_{\gamma_N}\vdots$	$\vdots$
invisible quantum $\downarrow$	$\vdots$	$\vdots$	$\vdots$	$\vdots$	$\vdots$

It is well known that Heisenberg had once complained of the absence of a universal constant with dimensionality of length in quantum theory in relation with the ultraviolet divergences in quantum field theory. Owing to this absence we cannot specify a *fixed* boundary (i.e., the so-called "Heisenberg cut") between quantum and classical, but, once a constant of this sort were present, such an absolute limitation would have been implied that the physical world smaller than certain lengh scale can never be visible owing to the absolute boundary between visible classical and invisible quantum worlds. In the above formulation based on the notion of sectors, we can, fortunately enough, benefit from two kinds of freedoms, one to specify a *temporary and conditional boundary* between classical and quantum regimes and the other to *shift* this boundary according to the relative configurations of length scales of factors relevant to the observed situation. Such a "conditional" boundary can clearly be defined mathematically *in a specified context of restricted degrees of accuracy*, as follows. Actually, this is in sharp contrast to the traditional approach where the absence (*) of *inter-sectorial* classical level in QM inevitably forces us to treat *separately either purely quantum or purely classical* situations, neglecting mutual connections and/or transitions between them! This lack of theoretical formulation always needs be made up by *heuristic* arguments, which may often be *wrong* as shown below.

By the above definition of sectors, any observables A of a quantum-classical composite system in a mixed phase $\pi = \bigoplus_{\gamma \in Sp(\mathfrak{Z}_\pi(\mathcal{A}))} \pi_\gamma$ with many disjoint sectors π_γ have the following *block-diagonal* structure:

$$\pi(A) = \begin{pmatrix} \pi_{\gamma_1}(A) & 0 & \cdots & 0 \\ 0 & \pi_{\gamma_2}(A) & 0 & \vdots \\ \vdots & 0 & \pi_{\gamma_3}(A) & 0 \\ 0 & \cdots & 0 & \ddots \end{pmatrix},$$

where all the off-diagonal terms can be shown to be 0 according to the mutual disjointness of π_γ's. Therefore, any expectation value $\langle\psi|A\psi\rangle$ in a *vector state* $\psi = c_1\psi_1 \oplus c_1\psi_2 \oplus \cdots \in \mathfrak{H}_\pi$ reduces automatically to a *mixed* state:

$$\begin{aligned} \langle\psi|\pi(A)\psi\rangle &= |c_1|^2 \langle\psi_1|\pi_{\gamma_1}(A)\psi_1\rangle + |c_2|^2 \langle\psi_2|\pi_{\gamma_2}(A)\psi_2\rangle + \cdots \\ &= Tr\rho_\psi\pi(A) \end{aligned}$$

with a *density operator* ρ_ψ given by:

$$\rho_\psi = |c_1|^2 |\psi_1\rangle\langle\psi_1| + |c_2|^2 |\psi_2\rangle\langle\psi_2| + \cdots.$$

Thus, such a common belief in QM is *wrong* that any vector state given as a *superposition* $c_1\psi_1 + c_2\psi_2 + \cdots$ is a pure state showing quantum *interference* effects. From this viewpoint, the famous paradox of Schrödinger's cat is merely an *ill-posed* question, based on the level confusions about *quantum-classical boundaries*. Namely, because of the absence of such a physical observable A as $\langle\psi_{\text{dead}}|A\psi_{\text{alive}}\rangle \neq 0$, the actual transition from the cat's being alive to dead can take place, not at the micro-level of the Geiger counter, but by *macroscopic accumulation* of infinitely many microscopic processes! This last point can be understood by such ***quantum-classical correspondence*** that *classical Macro* level consisting of order parameters to describe inter-sectorial structure *emerges* from microscopic levels *through condensation of infinite number of quanta*. (As a matter of course, the presence or absence of microscopic observables triggering macroscopic state changes depends highly on the situations and/or aspects in consideration, like the case, for instance, of the visual eyesight controlled by photo-chemical reations involving the rhodopsin molecules at the retinae.)

Thus, it is important to understand properly the roles played by a *systematic methodology to construct a unified system combining extreme cases into one*, where one of the most important advantages of the scheme based on "***Micro-Macro duality***" can be found: What we attain in a "unified" treatment is not only a simple unification of two (or more) poles but the implementation of *bi-directional transitions* among them, inherent in the very notion of ***duality.***

4. Fourier-Galois Duality & Bi-directionality $\Rightarrow$"*Inverse Problem*"

The most familiar sort of duality is well known in *Fourier duality*, which allows one to go back and forth between an abstract group G and its group dual $\hat{G}$ or category $RepG$ of (certain class of) representations of G. This can be extended to the ***Fourier-Galois duality*** as a duality between [a dynamical system consisting of an algebra $\mathcal{F} \underset{\tau}{\curvearrowleft} G$ acted on by a group G] and [the *fixed-point subalgebra* $\mathcal{F}^G \underset{\hat{\tau}}{\curvearrowleft} \hat{G}$ with a *co-action* $\hat{\tau}$ of G (= action of $\hat{G}$ or $RepG$)]: i.e.,

$$\begin{array}{ccccc} & [\mathcal{F} \underset{\tau}{\rtimes} G & \simeq \mathcal{F}^G] \underset{\hat{\tau}}{\curvearrowleft} \hat{G} & & \\ \begin{array}{c}\text{crossed}\\ \text{product}\end{array} : \rtimes G \nearrow & & & \swarrow \rtimes \hat{G} : \begin{array}{c}\text{crossed}\\ \text{product}\end{array} \\ \text{dyn.sys:} \quad G \underset{\tau}{\curvearrowright} [\mathcal{F} = & & \mathcal{F}^G \underset{\hat{\tau}}{\rtimes} \hat{G}] & & \end{array}$$

Its basic structure fits well to our quadrality scheme:

$$\begin{array}{ccc} & \textbf{\textit{Spec}}\text{: } \hat{G} & \underset{\hat{\tau}}{\curvearrowright} \\ \textbf{\textit{Alg}}\text{: } \mathcal{F} & \rightleftarrows / \nearrow / \swarrow / \rightleftarrows & \mathcal{F}^G\text{: } \textbf{\textit{Rep}} \;. \\ \underset{\tau}{\circlearrowleft} & \textbf{\textit{Dyn}}\text{: } G & \end{array}$$

A concrete and important example of this kind of duality can be found in Doplicher-Haag-Roberts (DHR) superselection theory with unbroken symmetry [5, 6]: in the direction $\leftarrow$ from Macro to Micro, the fixed-point subalgebra $\mathcal{A} = \mathcal{F}^G$ consisting of *G-invariant observables* can recover the group G of symmetry and the whole dynamical system $\mathcal{F} \underset{\tau}{\curvearrowleft} G$, respectively, as the ***Galois group*** $G = Gal(\mathcal{F}/\mathcal{A})$ and the ***Galois extension*** $\mathcal{F} = \mathcal{A} \underset{\hat{\tau}}{\rtimes} \hat{G}$ of $\mathcal{A}$ (= coefficients of the "equation"), when $\mathcal{A}$ is combined with phenomenological data on its *sector structure* parametrized by $\hat{G}$ (= solutions of certain "equation"); this is just the core of ***Fourier-Galois duality***. Its essence lies in the recovery of the whole system through the *re-mobilization* by $\hat{G}$ of G-invariant fixed-points $\mathcal{F}^G$, analogous to G recovered from $\hat{G}$ by its inverse Fourier transform $\widehat{(\hat{G})} \simeq G$.

What is more important is the guarantee of universality and of uniqueness of the recovered results, which is due to the ***matching condition*** [4] to restrict the focus of our interest to the aspects of G-dynamical systems, in a consistent way with "Duheme-Quine Paradox"! In the case of the above DHR theory, such a matching condition can be found in the form of DHR selection criterion [5] to pick up all the states ω carrying localizable charges whose deviations from the vacuum ω_0 are restricted to finitely extended spacetime regions $\mathcal{O}$: $\pi_\omega \upharpoonright_{\mathcal{O}'} \cong \pi_{\omega_0} \upharpoonright_{\mathcal{O}'}$, where π_ω and $\pi_\omega \upharpoonright_{\mathcal{O}'}$ are the GNS representation corresponding to ω and its restriction to the local subalgebras in the causal complement $\mathcal{O}'$ of $\mathcal{O}$. While this theory originally applied only to the unbroken internal symmetries, it has been extended to the situations with spontaneously [4] and explicitly broken symmetries [7] and also to a general scheme for quantum measurements where measurement processes involving transitions from Micro to Macro is shown to be dual to the reconstruction of the original quantum system from measured data [3] in the opposite direction, from Macro to Micro. These examples of duality suggest the possibility to consider such problems as:

solving equations to obtain solutions $\rightleftarrows$ ***reconstructing equations*** from solutions,

duality between ***deduction*** (from Micro theory to Macro data) $\rightleftarrows$ ***induction*** (from Macro data to Micro theory), etc., etc.

In the search for a new structure to incorporate the old and standard one as a special case, one usually attempts trial-and-error searches in a heuristic way which seems to be unavoidable. How and to which extent can this be made systematic by the method for solving an "***inverse problem***"?

5. Tsallis Entropy & Scale Transformations

What I have so far elaborated on is the merits of examining the mutual relations among different levels of nature on the basis of Micro-Macro duality. Along this line, it would be interesting to examine the "*inverse problem*" to trace back a path specified by a sequence S_q of Tsallis entropy from the common knowledge of its limit point $S = \lim_{q\to 1} S_q$ as Shannon entropy:

	Tsallis	$\overset{??}{\dashleftarrow}$ $\underset{q\to 1}{\longrightarrow}$	Shannon
entropy	$S_q(\{p_i\}) = -\frac{\sum_{i=1}^{k} p_i^q - 1}{q-1}$		$S(\{p_i\}) = -\sum_{i=1}^{k} p_i \log p_i$
quantum entropy	$S_q(\rho) = -\frac{Tr\rho^q - 1}{q-1}$		$S(\rho) = -Tr\rho\log\rho$
relative entropy	$S_q(\rho\vert\sigma) = \frac{Tr\rho^q\sigma^{1-q} - 1}{q-1}$ $= qD^{(1-2q)}(\rho\Vert\sigma)$		$S(\rho\vert\sigma) =$ $Tr\rho(\log\rho - \log\sigma)$

.

Together with the corresponding quantum version, the relative entropy or the Kullback-Leibler information, as a quantity naturally related with the entropy, is given on the bottom line for two density matrices ρ and σ, in terms of which classical as well as quantum systems (with finite degrees of freedom) can be treated on the same footing. In contrast to the axiomatic-type discussions on non-extensive entropies [2], the focus point here is the mutual relation between the standard Shannon entropy S and the non-extensive Tsallis entropy S_q (as well as their relative versions), and the more important problem is *in which context* the paths $\underset{q\to 1}{\longrightarrow}$ absorbed into a *fixed point* S at $q = 1$ can be reversed uniquely $\overset{??}{\dashleftarrow}$ into the family of Tsallis entropies S_q parametrized by q. We expect that, when this problem is solved, it will give a clear-cut answer to the question, under which condition and in what sense one is necessarily forced to use generalized statistics with Tsallis entropy.

To prepare a setting up for this consideration, we need to clarify what is the *variable dual* to the parameter q so as to specify the context to validate a duality structure relevant to the problem. For this purpose, we recall the crucial roles played by *scale transformations* in *moving from one scale region to another.* We can formulate a scheme where scale invariance can be treated as a broken symmetry accompanied by the so-called *power-law behaviours* as the polynomical corrections to the vanishing exponent $m \to 0$:

		Spec	
	$\exp(-m(-))$ $\swarrow\nearrow$	Macro	unclear Micro-Macro boundary
clear Micro-Macro boundary with $m \neq 0$	$\swarrow\nearrow$	$\searrow m \to 0$ x^N: power-law corrections $\swarrow N \to 0$	***broken*** scale invariace
Micro	$\swarrow\nearrow \log_a$	$\frac{1}{x}$ in	conformal inversion
Dyn			

In [7] the (inverse) temperature β is shown to appear as order parameter associated with this symmetry breaking. In view of (non-commutative) L^p-theory formulated in terms of the modular operators $\Delta_\beta^{1/p}$ which can roughly be viewed as the Boltzmann factor $\exp(-\beta H/p) = \exp(-H(k_B pT)^{-1})$, this broken scale invariance can be related to the change of p appearing in L^p-$L^{p'}$ duality (with $1/p+1/p'=1$), and the latter is also related with the information geometry and α-divergence through such identification [7] as $1/p = (1-\alpha)/2 = q$ and $1/p' = (1+\alpha)/2 = 1-q$, or equivalently, $\alpha := 1/p - 1/p' (\neq \pm 1)$, $1/p + 1/p' = 1$. In this thermodynamic as well as interpolation-theoretical context, the meaning of Tsallis relative entropy can be naturally understood as the analytic continuation of the standard Shannon type relative entropy $S(\rho)$ given on the real-time axis $\mathbb{R}$ into the so-called KMS strip $\mathcal{D}_\beta := \{t + i\tau; t \in \mathbb{R}, 0 \leq \tau \leq \beta\}$ with *imaginary time* $i\tau$ where any thermal equilbrium states ω_β at the inverse temperature β are characterized mathematically by the KMS condition [8]. The formula for an α-divergence [9] is given in [7] for non-commutative L^p-spaces associated with an arbitrary von Neumann algebra $\mathcal{M}$ [10] by

$$D^{(\alpha)}(T_1||T_2) = pp'\left[\frac{||T_1||_p^p}{p} + \frac{||T_2||_{p'}^{p'}}{p'} - Re[T_1,T_2]_{\phi_0}\right],$$

where $[T_1,T_2]_{\phi_0}$ is the pairing between L^p and $L^{p'}$ w.r.t. a faithful normal semifinite weight ϕ_0 (or, a faithful normal state such as ω_β) for $T_1 = u_1\Delta_{\varphi_1,\phi_0}^{1/p} \in L^p(\mathcal{M},\phi_0)$, $T_2 = u_2\Delta_{\varphi_2,\phi_0}^{1/p'} \in L^{p'}(\mathcal{M},\phi_0)$, with u_i: partial isometries and $\Delta_{\varphi_i,\phi_0}$: relative modular operator from ϕ_0 to $\varphi_i \in \mathcal{M}_{*,+}$. This can easily be adapted to the above familiar case of density matrix states ρ and σ (typically identified with Gibbs states ω_β) by $T_1 = u_1\Delta_{\varphi_1,\phi_0}^{1/p} \to \rho^{1/p}$, $T_2 = u_2\Delta_{\varphi_2,\phi_0}^{1/p'} \to \sigma^{1/p'}$, $||T_1||_p^p \to \phi_0(|T_1|^p) = Tr((\rho^{1/p})^p) = Tr\rho = 1$, $||T_2||_{p'}^{p'} \to Tr\sigma = 1$, $Re[T_1,T_2]_{\phi_0} \to Tr(\rho^{1/p}\sigma^{1/p'}) = Tr(\rho^q\sigma^{1-q})$, and hence,

$$\begin{aligned} D^{(\alpha)}(\rho^{1/p}||\sigma^{1/p'}) &= pp'\left[\frac{1}{p} + \frac{1}{p'} - Tr(\rho^{1/p}\sigma^{1/p'})\right] = \frac{1 - Tr(\rho^q\sigma^{1-q})}{q(1-q)} \\ &= \frac{1}{q}S_q(\rho|\sigma). \end{aligned}$$

If we rewrite the expressions above by $D^{(\alpha)}(\rho^{1/p}||\sigma^{1/p'}) \to D^{(\alpha)}(\rho,\sigma)$, $S_q(\rho|\sigma) \to K_q(\rho|\sigma)$, we can easily see the agreement between the formulae here and those in [11]. Note that the GNS-representation space for density matrix states is given by the Hilbert space $\mathfrak{H}_{HS}$ consisting of Hilbert-Schmidt class operators with the square root $\rho^{1/2}$ of ρ playing the role of the GNS cyclic vector $|\Omega_\rho\rangle$. While the discussion here is given in terms of density matrices, this is simply for the sake of familiarity and the general essence can be easily extended to the systems with infinite degrees of freedom described by non-type I von Neumann algebras.

Now a *relative entropy functional* involving observables A and B can be introduced by generalizing Uhlmann's formulation [12] into the following form:

$$\begin{aligned} D^{(\alpha)}(\rho|\sigma)(A,B) \quad &: = \frac{1}{q(1-q)}\left[Tr(A^*\rho B) - Tr(A^*\rho^q B\sigma^{1-q})\right] \\ &= \frac{1}{q}\frac{\langle\Omega_\sigma|A^*(\Delta_{\rho,\sigma} - \Delta_{\rho,\sigma}^q)B\Omega_\sigma\rangle}{1-q}; \\ S(\rho|\sigma)(A,B) \quad &: = \overline{\lim_{q\nearrow 1}}\frac{\langle\Omega_\sigma|A^*(\Delta_{\rho,\sigma} - \Delta_{\rho,\sigma}^q)B\Omega_\sigma\rangle}{1-q} \\ &= \langle\Omega_\sigma|A^*\left(\Delta_{\rho,\sigma}\log\Delta_{\rho,\sigma}\right)B\Omega_\sigma\rangle = -\langle\Omega_\rho|B\log\Delta_{\sigma,\rho}A^*\Omega_\rho\rangle. \end{aligned}$$

While $Tr(\sigma(-)) = \langle\Omega_\sigma|(-)\Omega_\sigma\rangle = \omega_\sigma(-)$ expresses the procedure to take an expectation value, the appearance of ρ^q or the relative modular operator $\Delta_{\rho,\sigma}$ indicate the roles played by the so-called "escort distribution". In view of the appearance of the relative modular operators in various operator-convex functions, it would be convenient to unify all the relevant formulae in the form of the f-divergences or quasi-(relative)entropies [13]. In view of the close relationship among α-divergences, canonical correlations = Bogoliubov–Kubo-Mori inner products (BKM products), transport coefficients given in terms of BKM products [14] and the entropy production [15], the consideration of this sort would be useful not only for various problems related with *power-law behaviours* and also *conformal symmetry* encountered in many transient areas in nature, but also for statistical physical problems in the usual situations with valid additivity.

Along the above line of thought, the case of entropy can be treated *formally* by considering the equality $\rho = \Delta_{\rho,Tr}$ for a density matrix ρ and the corresponding relative modular operator between ω_ρ and the trace Tr (as a normal faithful semi-finite weight corresponding to $\sigma = 1$), which involves, however, the wrong sign due to the lack of normalization for the trace functional:

$$\begin{aligned} S_q(\rho; A, B) &= -\frac{Tr(A^*\rho B) - Tr(A^*\rho^q B)}{1-q} \\ \underset{A=B\to 1}{\to} \quad -\frac{Tr\rho^q - 1}{q-1} &= S_q(\rho) \underset{q\to 1}{\to} S(\rho). \end{aligned}$$

In any case, the above arguments show again that the origin of Tsallis entropy $S_q(\rho)$ can be found mathematically in the L^p-interpolation, whose physical essence lies in the analytic continuation of correlation functions into the imaginary time axis for describing the dissipative effects of thermal fluctuations. In short, $S_q(\rho)$ is just a *dilation* from the Shannon type entropy $S(\rho)$ living in the *real* world with real time t with vanishing imaginary time $\tau = 1/p = 0$ (corresponding to L^∞) into the dissipative domain $\tau \neq 0$ in the KMS strip $\mathcal{D}_\beta$.

Concerning the direction $\overset{??}{\dashleftarrow}$ itself opposite to the standard arrow $\underset{q\to 1}{\to}$, the context to be focused on is now clear in relation with the *interpolation theory* and its physical implications involving inverse temperature or *imaginary time* β. To solve this inverse problem in a literal sense, however, we need to proceed one step further. In this connection, it would be helpful to recall some more ingredients typically used in the context of q-deformed mathematics, among which the notion of *Jackson derivative* seems to be most relevant, as is suggested by Mr. H. Saigo: in the above discussions we see that the controlling parameter $1/p = q$ moves around in the imaginary time interval $[0, 1] \ni 1/p$ of the KMS strip $\mathcal{D}_\beta$, the former of which is *compact* (in the form of $1/p$ but may not be so in the form of $p \in [1, +\infty]$). It is clear that the controlling parameter for solving the inverse problem is *dual* to this, running over a *discrete set*, on which the Jackson derivative may act. In this way, many results obtained in the fields involving Tsallis entropy and related objects are expected to shed new lights on the above well-established traditional areas, while the effects acting in the opposite direction may also be beneficial for both sides. Indeed, this kind of *bi-directionality* is to be emphasized again as the core essence of the approach based on *Micro-Macro duality!*

Acknowledgments

Let me express my sincere thanks to Profs. H. Suyari, T. Wada and A. Ohara for inviting me to this inspiring workshop and for the continuation of fruitful discussions; I benefited so much from the lectures by and the discussions with Prof. Tsallis, the pioneer of this field, Prof. Topsoe, Prof. Naudts and Prof. Amari, and others. Last but not least, many discussions with Mr. H. Saigo, including his suggestion on the Jackson derivative are gratefully acknowledged.

[1] Tsallis C, *Introduction to Nonextensive Statistical Mechanics: Approaching a Complex World*, Springer (2009).
[2] Suyari H. Generalization of Shannon-Khinchin axioms to nonextensive systems and the uniqueness theorem for the nonextensive entropy, IEEE Trans. Inform. Theory, **50**, 1783-1787 (2004); Suyari H and Tsukada M, Law of error in Tsallis statistics, IEEE Trans. Inform. Theory, vol.51, 753-757 (2005); Suyari H. Mathematical structure derived from the q-multinomial coefficient in Tsallis statistics, Physica **A368**, 63-82 (2006); Wada T and Suyari H, A two-parameter generalization of Shannon–Khinchin axioms and the uniqueness theorem, Phys. Lett. **A368**, 199-205 (2007).
[3] Ojima I, Micro-Macro duality in quantum physics, 143-161, Proc. Intern. Conf. "*Stochastic Analysis: Classical and Quantum*", World Sci. (2005); Ojima I and Takeori M, How to observe and recover quantum fields from observational data? –Takesaki duality as a Micro-Macro duality–, Open Sys. Inform. Dyn. **14**, 307 - 318 (2007).
[4] Ojima I, A unified scheme for generalized sectors based on selection criteria –Order parameters of symmetries and of thermal situations and physical meanings of classifying categorical adjunctions–, Open Sys. Inform. Dyn. **10**, 235-279 (2003).
[5] Doplicher S, Haag R and Roberts J E, Fields, observables and gauge transformations I & II, Comm. Math. Phys. **13**, 1-23 (1969); **15,** 173-200 (1969); Local observables and particle statistics I & II, **23**, 199-230 (1971); **35**, 49-85 (1974).
[6] Doplicher S and Roberts J E, Why there is a field algebra with a compact gauge group describing the superselection structure in particle physics, Comm. Math. Phys. **131**, 51-107 (1990); Endomorphism of C*-algebras, cross products and duality for compact groups, Ann. Math. **130**, 75-119 (1989); A new duality theory for compact groups, Inventiones Math. **98**, 157-218 (1989).
[7] Ojima I, Temperature as order parameter of broken scale invariance, Publ. RIMS (Kyoto Univ.) **40**, 731-756 (2004).
[8] Bratteli O and Robinson D W, *Operator Algebras and Quantum Statistical Mechanics*, vol. 2 (2nd ed.), Springer-Verlag (1997).
[9] Amari S and Nagaoka H, *Methods of Information Geometry*, Translations of mathematical monographs; v. 191, Amer. Math. Soc. & Oxford Univ. Press (2000).
[10] Araki H and Masuda T, Positive cones and L_p-spaces for von Neumann algebra, Publ. RIMS, Kyoto Umv., **18**, 759-831 (1982); Masuda T, L_p-spaces for von Neumann algebra with reference to a faithful normal semifinite weight, Publ. RIMS, Kyoto Univ. **19**, 673-727 (1983).
[11] Ohara A, Geometry of distributions associated with Tsallis statistics and properties of relative entropy minimization, Phys. Lett. **A 370**, 184–193 (2007).
[12] Uhlmann H, Relative entropy and the Wigner-Yanase-Dyson-Lieb concavity in interpolation theory, Comm. Math. Phys. **54**, 21-32 (1977).
[13] Csiszár I, Information type measure of difference of probability distributions and indirect observations, Studia Sci. Math. Hungar. **2**, 299-318 (1967); Petz D, Quasi-entropies for states of a von Neumann algebra, Publ. RIMS, Kyoto Umv., **21**, 787-800 (1985).
[14] Kubo R, Toda M, and Hashitsume N, *Statistical Physics*, vol. 2 (Springer-Verlag, Berlin, 1985); Bratteli O and Robinson D W, *Operator Algebras and Quantum Statistical Mechanics*, Vol.2, Springer-Verlag. (1981).
[15] Ojima I, Hasegawa H and Ichiyanagi M, Entropy production and its positivity in nonlinear response theory of quantum dynamical systems, J. Stat. Phys. **50** (No.3/4), 633-655 (1988); Ojima I, Entropy production and nonequilibrium stationarity in quantum dynamical systems: Physical meaning of van Hove limit, J. Stat. Phys. **56** (No.1/2), 203-226 (1989); Ojima I, Entropy production and nonequilibrium stationarity in quantum dynamical systems, pp.164-178 *in* Proceedings of International Workshop on Quantum Aspects of Optical Communications (Nov.1990, Paris, France), Lecture Notes in Physics No.378, Springer-Verlag, Berlin-Heidelberg-New York, 1991.

Journal of Physics: Conference Series **201** (2010) 012018 doi:10.1088/1742-6596/201/1/012018

Maximization of Tsallis entropy in the combinatorial formulation

Hiroki Suyari

Graduate School of Advanced Integration Science, Chiba University, Chiba-shi, Chiba, 263-8522, Japan

E-mail: `suyari@faculty.chiba-u.jp`, `suyari@ieee.org`

Abstract. This paper presents the mathematical reformulation for maximization of Tsallis entropy S_q in the combinatorial sense. More concretely, we generalize the original derivation of Maxwell-Boltzmann distribution law to Tsallis statistics by means of the corresponding generalized multinomial coefficient. Our results reveal that maximization of S_{2-q} under the usual expectation or S_q under q-average using the escort expectation are naturally derived from the combinatorial formulations for Tsallis statistics with respective combinatorial dualities, that is, one for additive duality and the other for multiplicative duality.

1. Introduction

Since the introduction of Tsallis entropy[1] in 1988 for a generalization of Boltzmann-Gibbs statistics (BG statistics), the maximum entropy principle has been the main approach along the Jaynes' original idea[2]. Jaynes developed Shannon's monumental achievement as pioneer of information theory in 1948 for the new method nowadays called maximum entropy principle (MEP) in mathematical physics. Originally, Jaynes applied the maximization of Shannon entropy under a constant expectation to the mathematical reformulation of Boltzmann-Gibbs statistics. On the other hand, the standard and well accepted introduction for Boltzmann-Gibbs statistics has been to count the number of accessible microstates[3]. From the traditional point of view in statistical physics, the counterpart for the number of accessible microstates in Tsallis statistics has been missing. For this purpose, the q-multinomial coefficient was introduced from the mathematics of the q-exponential and it was shown to have a natural generalization of the one-to-one correspondence between the usual multinomial coefficient and Shannon entropy[4]. This relation also reveals the additive duality ($q \leftrightarrow 2-q$) in Tsallis statistics. After that, the q-multinomial coefficient was generalized to cover the four typical mathematical structures: the additive duality ($q \leftrightarrow 2-q$), the multiplicative duality ($q \leftrightarrow 1/q$), the q-triplet and the multifractal-triplet[5]. Based on these mathematical results, the maximization of Tsallis entropy is reformulated in the combinatorial sense.

This paper consists of 4 sections including this introduction. In order to clarify the essential idea, an important and simple example of Jaynes' MEP for Boltzmann-Gibbs statistics is first given in the section 2. In the following section 3, the MEP for Tsallis statistics is reformulated using the above combinatorial formula. The section 4 concludes the paper.

Journal of Physics: Conference Series **201** (2010) 012018 doi:10.1088/1742-6596/201/1/012018

2. A simple example of Jaynes' MEP for BG statistics

The present example was originally used by Boltzmann and later simplified in [6]. In order to clarify the mathematical formulations for MEP in Tsallis statistics from the combinatorial point of view, this example is generalized in the next section.

Suppose that n dice are thrown on the table and the total number of spots showing is nU_1. What proportion of the dice are showing face $i, i = 1, \cdots, 6$?

Let n_i be the number of the spots showing face i when n dice are thrown. Then there are $\begin{bmatrix} & n & \\ n_1 & \cdots & n_6 \end{bmatrix}$ such ways. This is a macrostate labeled by $(n_1, \cdots, n_6)$ corresponding to $\begin{bmatrix} & n & \\ n_1 & \cdots & n_6 \end{bmatrix}$ microstates. Each microstate has a probability $\frac{1}{6^n}$ which directly comes from the principle of equiprobability. To find the most probable macrostate, we wish to maximize $\begin{bmatrix} & n & \\ n_1 & \cdots & n_6 \end{bmatrix}$ under the constraint observed on the total number of spots, $\sum_{i=1}^{6} in_i = nU_1$. This maximization problem is formulated as

$$\text{maximize} : \begin{bmatrix} & n & \\ n_1 & \cdots & n_6 \end{bmatrix} \tag{1}$$

$$\text{constraint} : \sum_{i=1}^{6} in_i = nU_1 \tag{2}$$

Using the Stirling's formula, the number of microstates $\begin{bmatrix} & n & \\ n_1 & \cdots & n_6 \end{bmatrix}$ is approximated to Shannon entropy in the following sense.

$$\ln \begin{bmatrix} & n & \\ n_1 & \cdots & n_6 \end{bmatrix} \simeq nS_1 \left(\frac{n_1}{n}, \cdots \frac{n_6}{n} \right) \tag{3}$$

Thus, the logarithm ln is a monotone increasing function and we define $p_i := \frac{n_i}{n}$, so that the above maximization problem is equivalent to the following form.

$$\text{maximize} : S_1 (p_1, \cdots p_6) \tag{4}$$

$$\text{constraint} : \sum_{i=1}^{6} ip_i = U_1 \tag{5}$$

The above equivalence was originally given by G.Wallis in 1962 [7] after the original paper of Jaynes [2]. This maximization problem is well known to be solved through the Lagrange multiplier method:

$$L := S_1 (p_1, \cdots p_6) - \alpha \left(\sum_{i=1}^{6} p_i - 1 \right) - \beta_1 \left(\sum_{i=1}^{6} ip_i - U_1 \right) \tag{6}$$

and its solution is given by

$$p_i^* = \frac{\exp(-\beta_1 i)}{Z} \qquad (i = 1, \cdots, 6) \tag{7}$$

where β_1 is chosen so that $\sum_{i=1}^{6} ip_i^* = U_1$. Thus, the most probable macrostate is $(np_1^*, \cdots, np_6^*)$ where $n_i^* = np_i^*$ dice show face i. In statistical physics, β_1 is the inverse temperature, i.e.,

$$\frac{\partial S_1}{\partial U_1} = \beta_1. \tag{8}$$

In the next section, the above equivalence between two maximization problems is generalized to Tsallis statistics in accordance with the same procedure as above.

3. Reformulation of Jaynes' MEP for Tsallis statistics in the combinatorial sense

3.1. Preliminaries

The mathematical structure in Tsallis statistics was recently found to be originated from the very fundamental nonlinear differential equation [8][9](See the appendix):

$$\frac{dy}{dx} = y^q. \tag{9}$$

The solution to the above equation is given by the so-called q-exponential:

$$y = \exp_q(x) := [1 + (1-q)\,x]^{\frac{1}{1-q}} \tag{10}$$

defined for $x \in \mathbb{R}$ satisfying $1 + (1-q)\,x > 0$. In the course of the above derivation, the inverse to the q-exponential is appeared.

$$\ln_q x := \frac{x^{1-q} - 1}{1-q} \tag{11}$$

which is called q-logarithm defined for $x > 0$.

Then the q-product $\otimes_q$ is introduced as a natural generalization of the exponential law in the following way:

$$\ln_q(x \otimes_q y) = \ln_q x + \ln_q y, \tag{12}$$

$$\exp_q(x+y) = \exp_q(x) \otimes_q \exp_q(y). \tag{13}$$

Thus, the q-product is concretely determined [10][11].

Definition 1 *(q-product) For any $x, y > 0$ satisfying $x^{1-q} + y^{1-q} - 1 > 0$,*

$$x \otimes_q y := \left[x^{1-q} + y^{1-q} - 1\right]^{\frac{1}{1-q}} \tag{14}$$

is the q-product.

By means of the q-product (14), the q-factorial is naturally defined in the following form.

Definition 2 *(q-factorial) For a natural number $n \in \mathbb{N}$ and $q > 0$, the q-factorial $n!_q$ is defined by*

$$n!_q := 1 \otimes_q \cdots \otimes_q n. \tag{15}$$

Thus, we concretely compute the q-Stirling's formula [4].

Proposition 3 *(q-Stirling's formula) Let $n!_q$ be the q-factorial defined by (15). The rough q-Stirling's formula $\ln_q(n!_q)$ is computed as follows:*

$$\ln_q(n!_q) = \begin{cases} \dfrac{n \ln_q n - n}{2-q} + O(\ln_q n) & \text{if } q \neq 2, \\ n - \ln n + O(1) & \text{if } q = 2. \end{cases} \tag{16}$$

The above rough q-Stirling's formula is obtained by the approximation:

$$\ln_q(n!_q) = \sum_{k=1}^{n} \ln_q k \simeq \int_1^n \ln_q x\, dx. \tag{17}$$

The rigorous derivation of the q-Stirling's formula is given in [4].

In order to define the q-multinomial coefficient, the q-ratio $\oslash_q$, the inverse operation to the q-product $\otimes_q$, is also needed and similarly defined from the requirements:

$$\ln_q (x \oslash_q y) = \ln_q x - \ln_q y, \tag{18}$$
$$\exp_q (x - y) = \exp_q (x) \oslash_q \exp_q (y). \tag{19}$$

Then, we define the q-ratio $\oslash_q$[10][11].

Definition 4 *(q-ratio) For any $x, y > 0$ satisfying $x^{1-q} - y^{1-q} + 1 > 0$,*

$$x \oslash_q y := \left[x^{1-q} - y^{1-q} + 1\right]^{\frac{1}{1-q}} \tag{20}$$

is the q-ratio.

We apply these formulations, q-product and q-ratio, to the definition of the q-multinomial coefficient [4].

Definition 5 *(q-multinomial coefficient) For $n = \sum_{i=1}^{k} n_i$ and $n_i \in \mathbb{N}$ $(i = 1, \cdots, k)$, the q-multinomial coefficient is defined by*

$$\left[\begin{array}{ccc} & n & \\ n_1 & \cdots & n_k \end{array}\right]_q := (n!_q) \oslash_q \left[(n_1!_q) \otimes_q \cdots \otimes_q (n_k!_q)\right]. \tag{21}$$

Based on these formulations, Tsallis entropy is derived as a natural generalization of the usual correspondence (3) [4].

Theorem 6 *When $n \in \mathbb{N}$ is sufficiently large, the q-logarithm of the q-multinomial coefficient coincides with Tsallis entropy (23) in the following correspondence:*

$$\ln_q \left[\begin{array}{ccc} & n & \\ n_1 & \cdots & n_k \end{array}\right]_q \simeq \begin{cases} \dfrac{n^{2-q}}{2-q} \cdot S_{2-q}\left(\dfrac{n_1}{n}, \cdots, \dfrac{n_k}{n}\right) & \text{if } q > 0,\ q \neq 2 \\ -S_1(n) + \sum\limits_{i=1}^{k} S_1(n_i) & \text{if } q = 2 \end{cases} \tag{22}$$

where S_q is Tsallis entropy:

$$S_q(p_1, \ldots, p_k) = \frac{1 - \sum\limits_{i=1}^{k} p_i^q}{q-1} \tag{23}$$

and $S_1(n) := S_1\left(\frac{1}{n}, \cdots \frac{1}{n}\right) = \ln n$.

See [4] for the details and its proofs.

Therefore, Tsallis entropy is found to be the information measure uniquely determined by the nonlinear differential equation (9) [9]. Moreover, the parameter q in Tsallis statistics coincides with the parameter q in the generalized dimension D_q for multifractal systems in the following sense [12]:

$$\exp\left(S_q^{\text{Rényi}}(p_i)\right) = \exp_q\left(S_q^{\text{Tsallis}}(p_i)\right) = \exp_{1/q}\left(S_{1/q}^{\text{Tsallis}}(P_j)\right) \simeq \varepsilon^{-D_q} \tag{24}$$

where $S_q^{\text{Rényi}}$ is the Rényi entropy:

$$S_q^{\text{Rényi}} := \frac{\ln \sum\limits_{i=1}^{k} p_i^q}{1-q}, \tag{25}$$

P_j is the so-called escort distribution defined by

$$P_j := \frac{p_j^q}{\sum\limits_{i=1}^{k} p_i^q} \quad (j = 1, \cdots, k), \tag{26}$$

and D_q is the generalized dimension:

$$D_q := \lim_{\varepsilon \to 0} \frac{1}{q-1} \frac{\ln \sum\limits_{i=1}^{k} p_i^q}{\ln \varepsilon}. \tag{27}$$

In the following subsections, the maximization problem in Tsallis statistics is reformulated in the combinatorial sense in accordance with the section 2.

3.2. Maximization of S_{2-q} under the normal expectation

In Tsallis statistics, the maximization of (1) under the constraint (2) is generalized in the following form:

$$\text{maximize} : \begin{bmatrix} & n & \\ n_1 & \cdots & n_6 \end{bmatrix}_q \tag{28}$$

$$\text{constraint} : \sum_{i=1}^{6} i n_i = n U_1 \tag{29}$$

The q-logarithm $\ln_q$ is also a monotone increasing function and we define $p_i := \frac{n_i}{n}$, so that the above maximization problem is equivalent to the following form.

$$\text{maximize} : S_{2-q}(p_1, \cdots p_6) \tag{30}$$

$$\text{constraint} : \sum_{i=1}^{6} i p_i = U_1 \tag{31}$$

Here we used the one-to-one correspondence (22). The maximization of S_{2-q} under the normal expectation (31) has been considered in some papers [13][14]. The thermodynamic relation is obtained as

$$\frac{\partial S_{2-q}}{\partial U_1} = \beta_1. \tag{32}$$

See [14] for the details.

Note that this β_1 is the same as β_1 in (8). This case is due to the additive duality $q \leftrightarrow 2-q$ in Tsallis statistics. The multiplicative duality $q \leftrightarrow 1/q$ is also well known in Tsallis statistics, which maximization problem is reformulated in the similar combinatorial sense from the slight generalization of the correspondence (22).

3.3. Maximization of S_q under the q-average (q-normalized expectation)

In order to find the multiplicative duality $q \leftrightarrow 1/q$ in the relation between a generalized multinomial coefficient and Tsallis entropy, some generalizations of the mathematical formulas in the subsection 3.1 are introduced in [5].

Definition 7 *((μ,ν)-factorial) For a natural number $n \in \mathbb{N}$ and $\mu,\nu \in \mathbb{R}$, the (μ,ν)-factorial $n!_{(\mu,\nu)}$ is defined by*

$$n!_{(\mu,\nu)} := 1^{\nu} \otimes_{\mu} 2^{\nu} \otimes_{\mu} \cdots \otimes_{\mu} n^{\nu}. \tag{33}$$

where $\nu \neq 0$.

Definition 8 *((μ,ν)-multinomial coefficient) For $n = \sum_{i=1}^{k} n_i$ and $n_i \in \mathbb{N}$ $(i = 1, \cdots, k)$, the (μ,ν)-multinomial coefficient is defined by*

$$\begin{bmatrix} & n & \\ n_1 & \cdots & n_k \end{bmatrix}_{(\mu,\nu)} := \left(n!_{(\mu,\nu)}\right) \oslash_{\mu} \left[\left(n_1!_{(\mu,\nu)}\right) \otimes_{\mu} \cdots \otimes_{\mu} \left(n_k!_{(\mu,\nu)}\right)\right]. \tag{34}$$

where $n!_{(\mu,\nu)}$ is the (μ,ν)-factorial defined in (33).

Proposition 9 *((μ,ν)-Stirling's formula) Let $n!_{(\mu,\nu)}$ be the (μ,ν)-factorial defined by (33). The (μ,ν)-Stirling's formula $\ln_{\mu}\left(n!_{(\mu,\nu)}\right)$ is computed as follows:*

$$\ln_{\mu}\left(n!_{(\mu,\nu)}\right) = \begin{cases} \dfrac{n \ln_{\mu} n^{\nu} - \nu n}{\nu(1-\mu)+1} + O(\ln_{\mu} n) & \text{if } \nu(1-\mu)+1 \neq 0, \\ \nu(n - \ln n) + O(1) & \text{if } \nu(1-\mu)+1 = 0. \end{cases} \tag{35}$$

This formula is computed by the approximation:

$$\ln_{\mu}\left(n!_{(\mu,\nu)}\right) = \sum_{k=1}^{n} \ln_{\mu} k^{\nu} \simeq \int_{1}^{n} \ln_{\mu} x^{\nu} dx. \tag{36}$$

Based on these results, we obtain the one-to-one correspondence between the (μ,ν)-multinomial coefficient and Tsallis entropy as follows.

Theorem 10 *When n is sufficiently large, the μ-logarithm of the (μ,ν)-multinomial coefficient coincides with Tsallis entropy (23) as follows:*

$$\frac{1}{\nu} \ln_{\mu} \begin{bmatrix} & n & \\ n_1 & \cdots & n_k \end{bmatrix}_{(\mu,\nu)} \simeq \begin{cases} \dfrac{n^q}{q} \cdot S_q\left(\dfrac{n_1}{n}, \cdots, \dfrac{n_k}{n}\right) & \text{if } q \neq 0 \\ -S_1(n) + \sum\limits_{i=1}^{k} S_1(n_i) & \text{if } q = 0 \end{cases} \tag{37}$$

where $\nu \neq 0$,

$$\nu(1-\mu)+1 = q, \tag{38}$$

S_q is Tsallis entropy (23) and $S_1(n) := S_1\left(\frac{1}{n}, \cdots \frac{1}{n}\right) = \ln n$.

See [5] for the details and its proofs.

The multiplicative duality $q \leftrightarrow 1/q$ in (37) is recovered when $\nu = q$. Then, μ is determined as $\mu = \frac{1}{q}$ from (38), so that we obtain

$$\ln_{\frac{1}{q}} \begin{bmatrix} & n & \\ n_1 & \cdots & n_k \end{bmatrix}_{\left(\frac{1}{q},q\right)} \simeq n^q \cdot S_q\left(\frac{n_1}{n}, \cdots, \frac{n_k}{n}\right) \tag{39}$$

which reveals the multiplicative duality "$q \leftrightarrow \frac{1}{q}$".

In order to give the combinatorial formula of the maximization of S_q under the constraint the q-average (q-normalized expectation) in [15], the generalized multinomial coefficient on the left side of (39) is rewritten by means of the "form" of the q-multinomial coefficient (21).

$$\begin{bmatrix} & n & \\ n_1 & \cdots & n_k \end{bmatrix}_{\left(\frac{1}{q},q\right)} = \begin{bmatrix} & n^q & \\ n_1^q & \cdots & n_k^q \end{bmatrix}_{\frac{1}{q}} \tag{40}$$

Note that $\begin{bmatrix} & n^q & \\ n_1^q & \cdots & n_k^q \end{bmatrix}_{\frac{1}{q}}$ does not satisfy the definition of (21) because $\sum_{i=1}^{k} n_i^q \neq n^q$,i.e., $\sum_{i=1}^{k} \left(\frac{n_i}{n}\right)^q \neq 1$[1]. Instead of this inconsistency involved in the formula (40), we apply the distribution determined from $\left(n_1^q, \cdots, n_k^q\right)$ to the constraint. Summarizing the maximization problem in combinatorial form, we have

$$\text{maximize}: \begin{bmatrix} & n & \\ n_1 & \cdots & n_6 \end{bmatrix}_{\left(\frac{1}{q},q\right)} \left(= \begin{bmatrix} & n^q & \\ n_1^q & \cdots & n_6^q \end{bmatrix}_{\frac{1}{q}} \right) \tag{41}$$

$$\text{constraint}: \sum_{i=1}^{6} i \frac{n_i^q}{\sum_{j=1}^{6} n_j^q} = U_q \tag{42}$$

where this U_q is slightly different from U_1 because the q-dependent expectation (q-average) is employed in the above maximization problem. Remark that the q-average (q-normalized expectation) does not depend on the summation $\sum_{i=1}^{k} n_i = n$. Using (37), the above maximization problem is equivalent to the following form.

$$\text{maximize}: S_q\left(p_1, \cdots, p_6\right) \tag{43}$$

$$\text{constraint}: \sum_{i=1}^{6} i \frac{p_i^q}{\sum_{j=1}^{6} p_j^q} = U_q \tag{44}$$

The thermodynamic relation has been already obtained as

$$\frac{\partial S_q}{\partial U_q} = \beta_q \left(\sum_{j=1}^{6} p_j^q \right) = \beta \tag{45}$$

where $\beta_q := \frac{q}{q+(1+\alpha)(1-q)}\beta$, α and β are the Lagrange multipliers. β_q is the generalized inverse temperature which coincide with $1/kT_q$ (T_q: the physical temperature [17]). See [18] for the details. In addition, Obviously, $\frac{\partial S_q}{\partial U_q} = \beta$ is a natural generalization of (8).

4. Conclusion

This paper presents the two combinatorial formalisms for the well-known maximum entropy problems in Tsallis statistics. One is the use of the usual expectation and the other is the q-average. More general combinatorial formalism for Tsallis entropy can be obtained if one uses the one-to-one correspondence (37). But the combinatorial meanings in the generalized multinomial coefficient are still missing, which is remained as a future work.

[1] This case naturally leads us to the maximization of S_q under the constraint $\sum_{i=1}^{k} i p_i^q = U$, which has been proposed in [16] and used for several years. Since the paper [15] in 1998, this constraint has never been applied for MEP in Tsallis statistics, because some requirements as expectation are missing. (e.g., $\sum_{i=1}^{k} p_i^q \neq 1$.)

Acknowledgments
The author acknowledge support by the Ministry of Education, Science, Sports and Culture, Grant-in-Aid for Scientific Research (B), 18300003, 2009.

Appendix

	Boltzmann-Gibbs statistics	Tsallis statistics
fundamental equation	$\frac{dy}{dx} = y$	$\frac{dy}{dx} = y^q$
fundamental function	exponential function	q-exponential function
information measure	Shannon entropy	Tsallis entropy
multiplication	$\times$ (product)(independence)	$\otimes_q$ (q-product)
q	$q = 1$	$q \neq 1$ (q in the generalized dimension D_q)
fundamental operator	differential operator $\frac{df}{dx} = \lim_{h\to 0} \frac{f(x+h)-f(x)}{h}$	Jackson's q-differential operator $\frac{d_q f}{d_q x} = \frac{f(qx)-f(x)}{(q-1)x}$
divergence	Kullback-Leibler divergence	α-divergence $q = \frac{1-\alpha}{2} \quad (\alpha \neq \pm 1)$

References

[1] C. Tsallis, Possible generalization of Boltzmann-Gibbs statistics, J. Stat. Phys. 52, 479-487, 1988.
[2] E.T. Jaynes, Information theory and statistical mechanics, Phys. Rev. 106, 620-630, 1957;.E.T. Jaynes, Information theory and statistical mechanics. II, Phys. Rev. 108, 171-190, 1957.
[3] R.C. Tolman, The Principles of Statistical Mechanics. Oxford: Clarendon Press. 1938. Reissued, 1979. New York: Dover.
[4] H. Suyari, M. Tsukada and Y. Uesaka, Mathematical structures derived from the q-product uniquely determined by Tsallis entropy, Proc. of the 2005 IEEE Inter. Sym. on Inform. Theory (2005IEEE-ISIT), pp.2364-2368, 2005.; H. Suyari, Mathematical structure derived from the q-multinomial coefficient in Tsallis statistics, Physica A, 368, 63-82, 2006.
[5] H. Suyari and T. Wada, Multiplicative duality, q-triplet and (μ, ν, q)-relation derived from the one-to-one correspondence between the (μ, ν)-multinomial coefficient and Tsallis entropy S_q, Physica A, 387, 71-83, 2008.
[6] T.M. Cover and J.A. Thomas, *Elements of information theory*, Wiley, New York, 1991, (2nd edition, 2006)
[7] E.T. Jaynes, Probability Theory, Cambridge Univ. Press 2003 p.351.
[8] C. Tsallis, Dynamical scenario for nonextensive statistical mechanics, Physica A, 340, 1-10, 2004.
[9] H. Suyari and T. Wada, Scaling property and Tsallis entropy derived from a fundamental nonlinear differential equation, Proc. of the 2006 Inter. Sym. on Inform. Theory and its Appli. (ISITA2006), pp.75-80, 2006. [LANL e-print cond-mat/0608007]
[10] L. Nivanen, A. Le Mehaute, Q.A. Wang, Generalized algebra within a nonextensive statistics, Rep. Math. Phys. 52, 437-444, 2003.
[11] E.P. Borges, A possible deformed algebra and calculus inspired in nonextensive thermostatistics, Physica A 340, 95-101, 2004.
[12] H. Suyari, in preparation for submission.; H. Suyari, New developments in the studies of Tsallis statistical mechanics, BUTSURI, vol.63, pp.450-454, 2008. (in Japanese)
[13] J. Naudts, Generalized thermostatistics based on deformed exponential and logarithmic functions, Physica A 340, 32-40, 2004.
[14] T. Wada and A.M. Scarfone, Connections between Tsallis' formalisms employing the standard linear average energy and ones employing the normalized q-average energy, Phys.Lett.A, 335, pp. 351-362, 2005.
[15] C. Tsallis, R.S. Mendes, A.R. Plastino, The role of constraints within generalized nonextensive statistics, Physica A, vol.261, pp.534-554, 1998.
[16] E.M.F. Curado, C. Tsallis, Generalized statistical-mechanics connection with thermodynamics, J.Phys.A 24, L69-L72, 1991.
[17] S. Abe, S. Martínez, F. Pennini and A. Plastino, Nonextensive thermodynamic relations, Phys. Lett. A, vol.281, pp.126-130, 2001.
[18] H. Suyari, The unique non self-referential q-canonical distribution and the physical temperature derived from the maximum entropy principle in Tsallis statistics, Prog. Theor. Phys. Suppl., vol.162, pp.79-86, 2006.

AUTHOR INDEX